Time Travel in Theory and Practice

Time Travel

in Theory and Practice

Dr. John Yates

Chandor, Goa, India
& London, England

Cover design by Dr. John Yates
Book design by Dr. John Yates

Dr. John Yates, M.Sc., Ph.D., MIPA
President & Director, Institute for Fundamental Studies
Visit my website at www.ifsgoa.com
Email me at uvscience[at]gmail.com

British Library Cataloguing-In-Publication (CIP) data
A catalogue record for this book is available from the British Library.

Published by I.F.S Foundation, London, Mumbai & Goa

ISBN: 978-0-9570434-1-1

Preface

I wrote the book because I had invented time travel and wanted to explain generally how it works.

Perhaps equally importantly, I wanted to indicate the philosophical implications of my discovery.

The work contains material from quite a lot of papers which have been published at reputable scientific conferences, in some cases have appeared independently in print and have usually appeared in one form or another on my blog and website. Details of sources are usually on my present blog and website at http://ttjohn.blogspot.com/ and http://www.ifsgoa.com/ . I had started and ran the "International Journal of Theoretical Physics" and the "Institute for Fundamental Studies". I would like to thank the many leading mathematicians and scientists, often Nobel prizewinners, who assisted with and gave their names to these ventures, in particular Professor David Bohm, for his brave and elegant attempt to rationalise and clarify quantum theory, and Professor Clive Kilmister, for his erudite and relevant historical studies on Eddington's 'Fundamental Theory'. Neither of these lucubrations are actually used in this work, but they did shed light on trials for the future.

The "Institute for Fundamental Studies" is still alive and kicking and its main site is at its mansion at Chandor on the Goan Riviera, but occasionally I can be found at its site in Vasai, India or even Fulham, London.

My research in the invention of time travel involved countless hours at the library, many interviews with professionals and trips to perhaps as many as 30 countries.

Specifically I mention with great thanks my help from Professor Tim Maudlin, Dr. John Rothwell, Professor John Gruzelier, Professor Alan Cowey, Dr. Vincent Walsh, Professor George Sudarshan, the late Dr. Jeffrey Gray and Prince Louis de Broglie. I also wish to thank the late Dr. Jeffrey Gray, Institute of Psychiatry, Maudsley Hospital for letting me do tests on his synaesthete patients. And there are so many other worthy, great and important people, for example at Sheffield, Oxford and Cambridge Universities, and UCL Institute of Neurology, Queen's Square London, who helped to create this great experience.

This is not the end ! There remains much great work to be done and any correspondence or views will be treated with great welcome and respect.

Just email uvscience[at]gmail.com if you wish to express your views. My present blog and website are at http://ttjohn.blogspot.com/ and http://www.ifsgoa.com/ .

Postscript: For what it is worth, here are some previous comments on this project. Some date back two years so are not necessarily current or applicable to all comments:

"I wish you every success" David Papineau , Kings College, London

"Good luck with your research!" David Chalmers, Australian National University

"Very radical" Tim Crane, Philosophy Professor, University College, London

Contents

1

Introduction

This work contains material from a number of the essays written about my time travel apparatuses, which allow travel backwards and forwards in time.

And yes, the apparatus and methods do work, but they carry a cost of time and effort.

The introduction here briefly sums up the history of the various developments of this patented process, given in more detail in subsequent chapters.

The results of Kornhuber and Deecke (1965), not to mention those of Libet, are now well known. However the Libet work appeared some time after I had been doing some theoretical research with Robin Gandy at the University of Manchester, after discussions on the philosophy of time with Arthur Prior, whose presentist views I liked and respected, but did not altogether agree with. He had in fact introduced me to Robin Gandy and a paper (Yates, 1968) arose as a result of this work. It became clear that simple computer artificial intelligence techniques might well predict elementary results like Newton's laws, but we were not happy with that, even though I did do it. It just did not seem to encompass a summary of the true impact in real terms of even Newton's universe, and at the time I commented in the original paper (Yates, 1968) on the then recent work of Mario Bunge (Bunge, 1967) which seems to provide a sort of conceptual bridge to the real world. Mario Bunge at the time had been selected by me as a member of the editorial board of a physics journal which I had started and run, because I had been enthusiastic about his quasi-philosophical approach to the problems of physics. I recall that at the time, philosophy was largely persona non grata in some physics journals, notably 'The Physical Review' when it was run by Sam Goudsmit, though Goudsmit certainly had broad views on the subject (Christensen, 2009). And, as I recall it, when a paper would not suit the alternative recommendation of "Annals of Physics", very occasionally we got it in the "International Journal of Theoretical Physics". And indeed, we finally even got George Gamow onto our editorial board very near the end of his time, when he was so drunk he could not stand up, a common situation even today for forward-looking Russians, and indeed scientists. philosophers and nowadays unemployed astronauts. And of course, Richard Feynman eventually published his first and indeed probably the very first significant paper on quantum computers in the "International Journal of Theoretical Physics".

I suppose we could say that the tradition of ignoring or losing philosophers has been continued by some, notably recently by Hawking and as before, such people so often genuinely do not seem to understand the sense of the importance of philosophy to physics, as we can see for example in many recent reviews, often seemingly written in fury, of a recent Hawking (2011) book, "The Grand Design". Indeed as Callender points out ""Perspectivalism" holds that there doesn't exist, even in principle, a single comprehensive theory of the universe. Instead, science offers many incomplete windows onto a common reality, one no more "true" than another. In Hawking's hands this position bleeds into an alarming anti-realism". Perhaps more importantly, as someone who sees neutral monism as showing a reasonable window on the truth, I am now very inclined to favor the reasonable perspectivism of Ronald Giere (Giere, 2006) - and this is only a very distant and irreproachable relation to the work of Hawking.

It did soon become very clear that simple university physics graduate style models for time were not going to be enough. Clearly the original Prior model made a start, but there did not then seem to be any immediately useful way to embody such ideas in the mathematics. Eventually, by the study of category theory, that fact that one model (the simple contemporary geometric model) would not be enough, but that only fully dawned on me many years later and after a study of Varela's work and of category theory. In the meantime however, I had built and patented my first working time machine using EEG equipment, a CRO, analogue computers, and the work of Gregory, Kornhuber and gestalt psychology after the fashion of Lewin and Leeper. A good and clearly confirmable proof of concept was going to be required, however, and in common with so many experiments involving human perception, this is by no means easy.

There has recently been a resurgence of interest in temporal consciousness as described in, for example the work of Dainton (2000, 2008, 2010) .

Dainton gives fairly good discussions and Dainton (2010) does point out that there are problems with approaches amounting roughly to either McTaggart's A series or his B series individually and without the other. I would take it that those who adopt at least some of these views can be, up to a point, some of "the six blind men of Indostan" in the poem by Saxe (e.g. as in Hertzberg, 1955), whose various ways of describing an elephant are really hopeless except when all descriptions are considered together. It turns out, from my lucubrations described in this book, that the use of both an A series representation and a B series representation may prove more rewarding than either alone.

Beyond my paper (Yates, 1968), I was delighted at the work of Kornhuber and Deeke (1965). Of course I knew of the work of Broad (1938) and of Smart (1955) and how both of these philosophers had held the view that the A and the B series could both be of value in describing time. I rapidly took the view that whilst both A and B views had merit simultaneously, they should not necessarily be conflated.

Kornhuber discovered that before a voluntary movement, there is a long build up of neural potential in the part of the cortex responsible for motor control. The brain begins to react as much as a second before the physical act is noted. This can be much longer in some cases, such as the case of a pianist preparing a run of notes. Kornhuber called this evidence the RP or 'readiness potential'. Kornhuber's results gave interesting problems to those considering consciousness, freewill, and ideas such as what we now call 'qualia' and the 'binding problem'. And to this day, problems relating to such factors and importantly to 'free will' and its ramifications, are still being considered by philosophers and others concerned with such matters. But from my standpoint

Kornhuber's work was a brilliant scientific proof and clarification of the importance of Prior's approach to the study of time and indeed to any other discussions of time which went well beyond the conventional B series approach of Einstein and even Godel. I saw that maybe such an approach could return to physics the 'soul' which Einstein and Newton had ripped out of it, a matter which now seemed to clearly impose problems to properly computerising the field of physical axiomatics which had seemed to open up so endlessly after the work of Robin Gandy's former mentor, the great Alan Turing.

And not only that ! The position was even confirmed by the mostly later work of Benjamin Libet, and his endlessly debated 'half-second' !

It had become clear that Einstein's simplistic descriptions of his relativistic 'observers', let alone Newton's earlier results, simply did not come anywhere describing what people actually do or observe - in terms close to ordinary modern physics. Even today, there are very few penetrating examples of attempting to do so, and such attempts (like that of Huggett (2010)) really do not seem to do the matter proper justice - one has to looks more to Giere (2006) etc. and to general philosophical descriptions. And recent work like that of Du (Zhang , 2011) should be able to live with this matter, for reasons I am describing throughout.

But Kornhuber's work at least gave us a basis for experiments in physics. The free will concerns of other writers never imposed in the same way once both A and B series were considered, and this is discussed briefly but frequently here. This is done throughout the book but perhaps particularly in Section 2.01 and Chapter 7.

Our earlier working experiments (Yates, 1984) used the real results and experimental methods of Gregory (1970), where he used visual techniques to observe astronomic observations - which related to signals from objects now in our own past (Yates (1984), and later we used a new way as in Chapter 4, which involved experimental philosophy, as described in that chapter). We had evolved the techniques of Gregory so they would work with electroencephalographic measurements in our own laboratories and with the results of Kornhuber et al. etc. and our own readings, and achieved a working time machine.

Some of these matters are discussed again in Section 6.01, in Chapter 7 and indeed throughout the book and further experiments are either proposed or being done.

At a later point we may use, at the Institute for Fundamental Studies, the Information Field Theory of Torsten Ensslin (Ensslin, 2009) and the help of Feynman diagrams to carry though some of the mathematical integration difficulties.

Attempts to develop a theory for this 'specious present' are given in more detail in Chapter 7 and Chapter 2. Methods of involving the unconscious mind using dreams are given in Chapter 7. Chapter 3 goes on to use these methods directly. Interestingly enough in Sections 6.08 and 7.10 , preliminary attempts to 'creep up on' qualia and consciousness involves the inclusion of emotion in such scenarios. It turns out that in evolutionary robotics, a topic we later have to turn to, emotion may be very important in the evolutionary process. (Gershenson 1999, Heylighen 2010, etc.)

In Chapter 5 we relate the Many Bubble Interpretation to quantum mechanics and find that it gives a more down to earth understanding of quantum mechanics than the Copenhagen interpretation. This may ultimately lead to some further interaction with quantum theory, but that was largely unexpected and unintentional - and that in itself is often a feature or a good model or description.

Obviously modern mathematics is obsessed with the Leibniz interpretations (Chapters 6 & 7 and Section 2.01) and is unfortunately marred by the now largely unacceptable theological and clearly Panglossian implications of Leibniz's work (Chapter 2.01) but Leibniz probably couldn't have helped this because of the spirit of the times. This had all become abundantly clear to me from a reading of Margolis's work (Sections 6.03, 7.02 & 7.10) and how so many of his arguments could be used in other historic situations, in this case with reference to Leibniz, and the Lutheran religion with its emphasis on 'hard determinism', instead of the Roman Catholic Church as in Margolis's case. Indeed we see Newton's mathematics to be of a high character but largely we do not relate this mathematics to Newton's ideas on religion, alchemy and the like. But Leibniz, even today, sounds like a reasonable man and most present day mathematicians - it seems - do not realise the importance of distancing themselves from his metaphysical leanings. The net result is perhaps that there is not in existence the obvious mathematical structure to outline the MBI very quickly in terms of mathematics so we have to do as best we can.

Obviously, a consciousness simply lagging about half a second behind reality could make it appear epiphenomenal, and perception could be regarded as being a series of disconnected events, too late for action.

On that latter note Penrose had suggested that the brain sends unconscious quantum information backward through time. In the quantum world, time is symmetrical, or bidirectional (as it also appears to be in unconscious dreams). Aharonov and Vaidman proposed that quantum-state reductions send quantum information backward in time; backward time referral is the only apparent explanation for experimentally observed EPR effects in quantum entanglement. Further Matte Blanco concluded: ". . . the processes of the unconscious . . . are not ordered in time". All this is discussed in Tuszynski (2006). I sincerely doubt whether such views can readily be used in the ways these writers presently suggested but the detailed use of dreamwork and its different 'logic' is something which can be and has been applied immediately to the MBI.

It is plain that there is still doubt among scientists (Edelman, 2011) in the field as to whether or to what extent consciousness is causal. Edelman maintains for example that "It is the neural structures underlying conscious experience that are causal. The conscious individual can therefore be described as responding to a causal illusion, one that is an entailed evolutionary outcome of selection for animals able to make plans involving multiple discriminations". It is not crystal clear to me as to the significance of such comments, but the mute historical hand of Leibniz's view seems to be overshadowing and perhaps clouding any likely contingent mathematics..

Even with all these difficulties, historical and otherwise, we have tried to present a viable mathematical and experimental form for the Many Bubble Interpretation, or MBI. We have already seen that the MBI can deal with quantum theory up to a point at least without conflation. It was, however, intended primarily for use with the A series or at worst for a mapping of some sort of the A series onto the B series. Details are further supplied of the MBI throughout and particularly in Chapter 5, supplementing those already given in Chapter 7 and elsewhere.

What proof do we have that time travel occurs here ?

1. The general result that it does fit in with what we have done.

2. Resolution of the Andromeda paradox (Chapter 2.03)

3. The results from dream work (Chapter 3 and elsewhere)

4, The results from experimental philosophy (Chapter 4 , Chapter 2.01 and elsewhere), using the ideas of Liberman, Trope, M.K. Johnson and others.

5. My earliest results using Libet's work which led to the patent. We have done a lot of work on the initial Libet results. The work of Trevena was only one of the first and earliest attempts to improve on Libet, and the current work of Bill Banks (a helpful fellow traveller on the path to scientific progress but now sadly deceased) and the even more imposing work of Banks and Isham (Chapter 6.01, 6.02 and elsewhere), clearly go much further along that route, It may be said (wrongly) that we 'jumped the shark' in 1980 by using an analog computer and a CRO to obtain actual future predictions using the Kornhuber work, as mentioned above. And we accurately predicted, over short times, the outcomes of independent random number generators ! Halcyon days, but much more work is needed. We are continually reviewing this work and tried for a further grant last year but stand by the basic results.

6. We have a theory, the MBI theory, which can explain quantum mechanics better than Copenhagen and at the same time, solves the problems of time which multiverse theories like those of Everett claim to solve, but which are generally without good evidence for all their extremely large multiverse claims. As for the multiverse, it is just that the grey beards 'generally believe it to be so', When I discussed Everett's work with Sir Rudolf Peierls many years ago, something of that kind was the best available but by then I had not got to my MBI. Our current results are based on known qualities of the mind, and have produced measurable results.

How can we take the work further ?

Plenty of ways. Here are just three routes to be going on with.

1. We are applying hypnosis, particularly to synaesthesic cases, taking the Mondrian results of Kosslyn further (Yates, 2011)

2. The evolutionary robotics results using simple Braitenberg vehicles as we begin to describe in Sections 2.02 and 2.04 . And further more ambitious projects are discussed in Section 2.06.

3. We have long planned to do some more work on the Libet style experiments, having discussed the problems recently with Bill Banks, and may

still do so.

Since we are superseding quantum theory and general relativity theory in this work, there is still a long way to go and a lot of experiments to be done before easy physical transfers of individuals through time becomes straightforward. Many of the existing results in quantum theory and general relativity will remain standing after a fashion, so there is little to worry about in that direction. But queries about whether the ultimate answer we are seeking is something like the much quoted '42' should be removed in the long term, and without any similar need to reduce the explanation to traditional mathematical physics. That is to say, seeking merely numerical, algebraic or even other 'mathematical' answers to questions of meaning is likely to merely be part of the problem, rather than part of the solution. So if we can remove, avoid, or otherwise circumvent that part of the problem we are on the way to the answer. And I both hope and believe that my present writings place us in this direction. Of course, in some alternative approach, we could all become, like Leibniz, Lutherans or "hard determinists" of a sort, and assume that our behaviour is predestined. Lutheranism is not a bad social faith at all, and as such, I quite like it, but it is hardly a philosophy of physics. Such faiths or beliefs as Lutheranism seem to be a significant alternative - but I prefer science for our present ends.

And of course just email uvscience[at]gmail.com if you wish to express your views. As already stated my present blog and website are at http://ttjohn.blogspot.com/ and http://www.ifsgoa.com/

References

Broad C.D., (1938), "An examination of McTaggart's Philosophy", Vol II, Part I, Cambridge University Press

Bunge, M. (1967), Reviews of Modern Physics, 39, 463

Christensen T.M. (2009), "John Archibald Wheeler: A Study of Mentoring in Modern Physics", Oregon State University, Publication 3376778

Dainton, B.F. (2000), "Stream of Consciousness: Unity and Continuity in Conscious Experience". London: Routledge, International Library of Philosophy. ISBN: 0-415-22382-2

Dainton B.F., (2008), "The Phenomenal Self", OUP, ISBN: 978–0–19–928884–7

Dainton, Barry, (2010), "Temporal Consciousness", The Stanford Encyclo--pedia of Philosophy (Fall 2010 Edition), Edward N. Zalta (ed.), URL = <http://plato.stanford.edu/archives/fall2010/entries/consciousness-temporal/>.

Edelman G. M., (2011) "Biology of consciousness." Frontiers in Consciousness Research 2 (2011): 4.

Ensslin T.A.,Frommert M., Kitaura F.S.,(2009), "Information field theory for cosmological perturbation reconstruction and nonlinear signal analysis",Phys. Rev. D 80, 105005

Gershenson, C. (1999). Modelling Emotions with Multidimensional Logic. *Proceedings of the 18th International Conference of the North American Fuzzy Information Processing Society (NAFIPS '99)*, pp. 42-46. New York City, NY.

Gregory R.L. (1970),"The Intelligent Eye", Appendix B, 170 et seq., Weidenfeld & Nicholson, London

Giere R.N., (2006), "Scientific Perspectivism", University of Chicago Press, ISBN 0226292126

Hawking S., (2011), "The Grand Design", examples of book reviews at http://www.canada.com/story_print.html?id=b7c1e4a9-4faf-4261-9862-770f334416ee&sponsor=   , http://www.nytimes.com/2010/09/08/books/08book.html

Hertzberg H. T. E. (1955), "Some Contributions of Applied Physical Anthropology to Human Engineering." Annals of the New York Academy of Sciences 63, no. 4 (November 1, 1955): 616-629.

Heylighen F., (2010) "Cognitive Systems: a cybernetic perspective on the new science of the mind", Lecture Notes 2009-2010, ECCO: Evolution, Complexity and Cognition - Vrije Universiteit Brussel

Huggett N., (2010), "Evererywhere and Everywhen", Chapter11 p116 et seq, Oxford University Press.

Kornhuber, H.H.; Deecke, L. (1965). Hirnpotentialänderungen bei Willkürbewegungen und passiven Bewegungen des Menschen: Bereitschaftspotential und reafferente Potentiale. Pflügers Arch 284: 1–17 "Citation Classic"

Smart J.J.C., (1955), "Spatialising Time", Mind, 64

Tuszynski J.A., (Ed.), (2006), "The Emerging Physics of Consciousness", p204, Springer , ISSN 1612-3018

Yates J., (1968), International Journal of Theoretical Physics, 171, Vol1, No 2.

Yates J., (1984) Patent Number:GB2051465 Publication date:1981-01-14. I also mention and apply Gott's comment to this patent in http://philpapers.org/archive/YATASO.1.pdf

Yates J., (2011) Several published papers reprinted in this volume refer to this work, including those in Sections 5.04, 7.02 and 7.10

Zhang, S., Chen J., Liu C., Loy M., Wong G., Du S., (2011), "Optical Precursor of a Single Photon." Physical Review Letters 106, no. 24 (June 2011).

2

Important Features and Uses of the MBI

Section 2.01 puts right some of the philosophical problems which modern science labours under.

Sections 2.02 & 2.04 develop and build a physical model (out of Lego parts in the case dealt with) which helps to describe the so-called 'specious present' and other related natural phenomena.

Section 2.03 resolves the Andromeda Paradox using the MBI.

Section 2.05 gives various ways and means which can relate brain activities to 'acceptable' physical phenomena.

Section 2.06 suggests the use of biocomputers and portrays conceivable future scenarios.

2.01 Not Even Wrong - a view of current science of the mind

Abstract: Present progress in mind science is racing away in the direction of denying the existence of human freewill and animal and human sentience. This brief paper attempts to summarize a few brief reasons why areas of present work by prominent authors have departed from fact to the realms of folk psychology and summarises some of the ways in which present work can be put right. An experiment is described and carried out in an attempt to breach a little more of the present gap between experimental fact and the outmoded theory which others have tried to apply blindly.

"We, the Party, control all records, and therefore we control all memories. Then we control the past, do we not ?" Big Brother (from George Orwell, "1984"). "Myths which are believed in tend to become true", George Orwell.

" Think for yourselves and ensure others enjoy the privilege to do so, too", Voltaire .

Introduction

Firstly, three basic aspects of mind and consciousness problems are briefly discussed and conclusions given as to measures to be adopted [Note 1].

(1) Overall philosophy of approach

Whilst many general descriptions of what goes on in the mind would appear to assume mathematics is the key to understanding the workings of the mind, this seems to be based at least in part on the assumption of some form of Platonism or something of the same stripe. In summary James Jeans said "the universe appeared to have been designed by a pure mathematician", and that it is "more like a great thought than a great machine".

This sort of thing is simply naive folk psychology.

There are many reasons why this variety of Platonism is simply folk psychology. I will not try to give all of them, as general statements of Jeans's sort have to have the relevance of their meaning proved, and this has not been done. Feynman [4] bluntly pointed out "We come across these mathematical relationships but they apply to the universe, so the problem of where they came from is doubly confusing....Those are philosophical problems I don't know how to answer".

Another simple and related reason is that if the universe has limits - and if it is assumed to be infinite this brings a further host of queries which certainly do not propose to support the above Jeans type folk psychology [Note 3]. and if the speed of light is finite, then our horizon only supports a limited number of particles, often computed as about 10^{80} . So even if the whole universe were to act as some kind of cosmic computer, it is only of a defined size. And because the universe is believed to be expanding, the resources inside our horizon are time-dependent and limited, ruling out various large numbers calculations as unpredictable. And mathematics containing an infinite number of steps could never be carried out. Further, mathematical results are time-dependent because of the variation in the number of particles. Indeed if there were a 'big bang' at the start , there would have been only a small number of particles then with a tiny computing power [49].

As Kant pointed out in a similar context "if we look through rose-tinted spectacles it is no wonder the world looks rosy". That is to say, to use concepts obtained through mathematics and computers as a groundwork for contemporary physical reasoning is simply current folk psychology, just as "ghoulies and ghosties and long-legged beasties, and things that go BOOMP in the night" helped to give us the Christian Litany of yesteryear. I even used to have a prayer book where God was seriously asked to deliver us from those things. Now people try to use mathematics in the same way as primitive people used God.

If mathematics works, up to a point, by analogy it only gives us the map and not the country.

Now there are serious wannabe 'new Platonists' like Roger Penrose [18] and their ideas are certainly worth examining. In substance they seem to claim that there is some deep underlying accord between Plato's world and the true physical world [50].

Their problem is that they never define that accord in useful detail, and they

Time Travel in Theory and Practice

cannot do so. But there seems no reason why we cannot consider that mathematics does give us some kind of map, and this would also explain psychologically the reason for the firmness of the views of those such as Penrose. Certainly, mathematics works, but whether it works 'surprisingly well', or simply works up to a point is purely a matter of conjecture and a social scientist or a biologist might well hold a different view to that of a physicist. For example in serious considerations of life itself, current mathematics has always appeared to be insufficient, as shown historically by Rashevsky and Rosen [5] for example.

Our present methods use the McTaggart B series for some of the mathematics, and use the A series as well for some parts of the work. B series mathematics alone does not describe the human situation completely and adequately and therefore is insufficient.

(2) Physical Problems for a B series only approach

Callender [6] points out, particularly in section 4 of his paper, that special relativity is inconsistent with any philosophically interesting conception of tense. In fact any notion of 'becoming' remotely similar to that found among advocates of the tensed view of time is not compatible with Minkowski spacetime. In my opinion that is exactly the sort of thing McTaggart's paradox is all about anyway. Since special relativity is necessary for the appropriate B series descriptions, the B series description alone will not suffice. Indeed to introduce the ideas like those of 'becoming' we would also need a further set of results and the A series falls into that category. In earlier works I repeatedly pointed out that an A series model is most unlikely, if not impossible, to be completely compatible with appropriate the B series model. Nonetheless both models can exist and a B series representation of the A series model can also be created.

It really depends on whether we want to give a complete description of time. Philosophers and scientists who do not, are like cartographers who claim that a two dimensional flat map shows heights adequately. Those who do, want to provide complete descriptions and to describe time as completely as they currently can, can use the A series as well as the B series, thus hopefully obtaining more information and enlightenment.

If 'becoming' and such ideas are truly not necessary in some parts of a particular program, fortunately they do not have to be included but it is unreasonable to leave them out at the start.

Other views on the McTaggart Paradox are of course currently common, and they are discussed in probably enough more detail for the present program in Appendices 2 and 3 and in the body of this work. Appendix 3 is a brief discussion with Tim Maudlin, who is perhaps the foremost expert on the McTaggart Paradox today.

(3) Human and nonhuman requirements

Most people believe that they have free will. The B series as usually devised does not allow free will to feature. In fact earlier writings in this series [7] make it plain that actual physical measurements [8, 15,16] using normal B series physics show that it certainly seems impossible to represent freewill properly in B series physics. It is not assumed here that people do have freewill, but the existence of the A series and its use, make free will a possibility in a universe world-view. If the possibility does not exist, the universe model does not allow free will to be confirmed or denied, making the model a very incomplete one. That is clearly true even if individuals choose to deny free will. For to permanently deny free will to every creature in the universe - probably including any potential Gods, extraterrestrials or any truly high powered intelligence - seems to run against Occam's razor and indeed normal common sense [Note 4].

Further, recent studies [19] suggest that undermining our everyday concept of free will can alter our ethical behavior, a very serious consequence indeed.

Interim Conclusions: As a result of many considerations including (1), (2) and (3) above, in further studies and particularly the present study, both the A series and the B series are considered. Hopefully the present approach will also be able to eventually heal the gap between the theories of Metzinger [11] and Noe [12], whose ideas in part seem almost diametrically opposed to one another but who use much the same experimental data [Note 2].

Introduction to Details of Experiment

From a practical point of view, the article in Wikipedia [63] suggests a number of philosophical, religious and other ways of approaching the problem of freewill. I do not consider this essay to be tied irretrievably to any of these, nor bound by any of them ! In short I propose not to meander in a purposeless philosophical jungle but simply to deal with the facts in a way as presented herein.

We use experimental philosophy techniques developed from the work of Marcia Johnson [2] and using similar and often identical queries to hers. The general philosophy of approach, though importantly not necessarily the specific assumptions, which we try to adopt is that of the very early work of Trope and Burnstein [64].

Construal Level Theory (CLT) [1] suggests that thinking about events that are far into the future or the past or considering any events which are remote, either psychologically or in some sense physically, and particularly events which seem unlikely or alternatives to reality, triggers a more global brain processing style. In analogue, it is like seeing the forest per se, and not the trees.

Rohrer [13] discusses how both our neural and developmental embodiment shape both our mental and linguistic categorizations. The degree of thought abstraction has been found to be associated with physical distance which then affects associated ideas and perception of risk.

Work like that of Grenander [14] on pattern theory can possibly be used eventually in a somewhat similar approach to ours, but we retain our earlier Berkeley Madonna models such as **N003b** [7] for the moment. Our models seem as if they could do with extending in ways which either use the A series directly, or a further extension within an A series model within the B series, the former perhaps becoming more and more necessary as further electrophysiological results become available. At all times we need to bear in mind less than optimistic appraisals like those of Hacker [32] ,Vul [33] . and nowadays many others. The difficult zone is probably the much over-hyped 'neuroeconomics' idea and psychological results on such work as the 'prisoner's dilemma' [34] and the like, which have to be more carefully evaluated than they have been. Looking carefully at popular books [35] like "The Newtonian casino" where the bulk of the hard work in experimentation seems to have been to disguise results from casino staff rather than to use such important new quantum methods as those of Doyne Farmer and Norman Packard [36] , and indeed the caustic comments of Dan Ariely [37] on the psychology of the recent banking crisis and the situation in banking as known to myself and many others for years, we can easily notice that so many modern methods are used in a way which are unfortunately self-serving mostly for the benefit of non scientific persons involved rather than precisely scientific in nature. This has to mean, at the very least, that a lot of care is necessary as self-serving practices can easily and even unintentionally obscure the scientific results and that is of nobody's interest in the long term.

For the moment, use of just the A series can only be done somewhat indirectly by using modern psychological techniques like Construal Level Theory (CLT) and being careful not to insist - especially unintentionally by implication - on the direct and necessary ultimate involvement of B series physics, probably as distinct from normal statistical methods such as Bayesian statistics.

Predicting the hedonic effect of a future event can be done by simulating it [3] and such facts immediately suggest bringing in the methods of CLT.

Olaf Blanke [10] wrote a very interesting paper on mental time travel (MTT), and Gilbert [17] wrote a review on a similar theme. It is certainly thought provoking and certain aspects of it mirror my own ideas. But Blanke, in his work on 'near death experience' for example, has been noted to jump to rather too obvious conclusions of the sort which seem he may well have missed a few steps in order to obtain credibility, and must examine his work fairly closely, not taking conclusions from correct experimental results necessarily on face value.

Our earlier experiments [20] on the examination [22, 69] of dreams do generally fit in with Construal level theory, especially in that the dreams contained unexpected elements of the future in an abstract form more often than in a very concrete form, although both occurred.

Clearly if we plan to invoke the A-series, it is easy to see how the elements of the future construed or envisaged in the mind of the present, have a good deal in common with the elements of the past. Rhyming philosopher/psychologist Alexander Pope many years ago stated "Remembrance and reflection - how allied ! What thin partitions sense from thought divide". Our understanding of memory today tries to stress "thick partitions" that divide sensory experience and thoughts or memories. This view considers that processing of sensory information and later cognitive activity can change thoughts and experienced memories. If one is prepared to accept at least the possibility of subscribing, at least up to a point, to this well researched and frequently accepted view, then if we look at contemplations of future events, the partitions between thoughts of past events and thoughts of future events seem as if they could be a lot thinner than many people nowadays try to suggest. A common approach in considering thoughts of the future is to revert to B-series physics and look for simple causal relationships between events as they occur. But this is simply an interpretation of what is being noticed. We are concerning ourselves with facts. The basic standpoint might be rather to take as first starting point a WYSIWYG viewpoint, in that we are measuring mental phenomena and there appears on the face of it, little basic difference in the state of mind of the individual between memory and forecast, though there is a tendency to pull down the shutters of the mind,as it were, and assume that "we cannot see the future" and so forth. I do not claim anything as simple as that, rather that we should thrust aside the shibboleth that only the B-series of Newton, Leibniz and for that matter Einstein is going to provide simplistic explanations of the universe. To immediately use fMRI results to justify the results of such as Leibniz, Newton and Einstein is really a circular argument. Most certainly, we should, as Addis [66] and many others have done, learn all we can about the brain, but it is important to take into account the conclusions of such people as Hacker [32] and Vul [33] . In fact we must go further and know that careful interpretation of fMRI results and so on may fall outside the realm of Newtonian physics. It seems probable that even in a fairly accurate A-series representation, the so-called 'future' and the 'past' have somewhat different configurations, but they certainly seem like each other. Differences have to be considered, but the most obvious problem is with ourselves, that most people always feel somewhat assured that we remember the past, and are less clear on the future. This of course tends to be borne out when we check but is possibly not part of the initial mental process. It is possible to think of mystical contemplation and other alleged things of this sort which often regard the human position in the universe as in some way timeless, but this idea tends to be a red herring in our present lucubrations, except insofar that these alleged phenomena at least provide a clue to the fact that modern (essentially Western) ways of thinking are a limited and very restricted way of looking at the world. The idea of assuming that any other approach simply requires oddball stimuli [70] (which are themselves an important consideration of course) or some such special situation, restricts current thinking of real phenomena too much.

The present experiment, therefore, looks to see if the results we obtain for an experiment somewhat like Johnson's are like those of Johnson [2] in waking time with similar or the same subjects as we previously used, and we bear in mind the fact that with modern CLT [21], increased temporal distance should increase the overall attractiveness of a high-level construal value relative to a low-level construal value. To quote [21] "A common assumption in the behavioral sciences is that the value of an outcome diminishes as temporal distance from the outcome increases - positive outcomes seem less positive when removed in time (intertemporal discounting). The prediction from CLT, however, is that increased temporal distance, as with any psychological distance, should shift the overall attractiveness of an outcome closer to its high level construal value and away from its low-level construal value. When the low-level value of an outcome is more positive than its high-level value, temporal discounting would obtain, so that the outcome would be less attractive in the more distant future. When the high-level value of an outcome is more positive, however, the outcome should be more attractive in the distant future .... thinking of trees may prompt us to think of tomorrow, whereas thinking of the forest may prompt us to think of next year. The link between distance and construal has important implications for perception, categorization, and inference".

Experimental Procedure and Results

The aim was not to produce important new confirmatory material in a sort of analogy to the Millikan oil drop experiment, which itself raised a great deal of controversy [72], but instead to see if and how any experiment in the A series could be designed. Hopefully, it may be even possible to allow such an experiment to begin to act as a prototype. Any useful result would simply be regarded as a plus point and a minor assistance as a minor proof of concept and to be a step on a way to provide an A series format. In the event, that is approximately what happened. Some confirmatory material is of course already available [20].

Details are given in Appendices 1a and 1b.

20 subjects, from the same group as was used for earlier experimentation [20], were used for 30 interviews using a total of approximately 24,000 queries.

The results agree with those of Trope [1] and many others, in that the degree of abstraction seems to increase with the time differential between the moment of the test and when the event is to conceived to take place.

We used the 1 to 7 scaling procedure of Johnson [2] and scoring was found accurate to about $\pm$ 1 at the 80% confidence level and $\pm$ 2 at the 98% confidence level. The percentage of results which could be construed as 'abstract', by occurring in the first quartile of an abstract-concrete scale were as follows. Distant Past 40% abstract, Past 17% abstract, Present/fantasy 33% abstract, Future 1% abstract, Distant Future 25% abstract. Also the correlation coefficient of degree of abstraction at the time of testing with both level of perceived detail and the level of personal involvement was high in the near past and distant future but lower in the distant past and near future .

Conclusions

Clearly this is only a beginning. It seems to me that a major difference in the present experiments is where we included future events as retaining a right to some kind of reality, as well as past events, present events and merely imagined events in our very simple survey, and we even attempted to begin to clearly distinguish presently imagined events from real past and future events. Reality monitoring [2] is of course essential as a guide to the relevance of such results.

We need many more experiments in experimental philosophy and it may need a survey device somewhat similar to the Amazon Mechanical Turk [65] and other such ideas to get a lot of results. Though such experiments might be inexpensive and realistic, they must at all times be hands-on, carefully planned and not mindlessly computerized

A further consideration or lemma is implied that not just the neural basis of memory and future must be considered in the way of Addis [66] , nor as the investigations of Trope [1] imply, but neural computation using neural computers of essentially an analog kind may be needed. I used analog computers in my very earliest experiments [67] and their use is in essence different to that of digital computers. After the many years of work by Minsky and many others on AI , it has become certain that simplistic digital computers are unlikely to do the whole job or even impossible to use effectively in the area. At least B-Z computers and similar devices may enhance progress [68].

Acknowledgments

I wish to thank Nandini Subramaniam Yates for organizing the experiments and for rendition of some of the material into English from Marathi and Hindi. My heartfelt thanks is also due to all the subjects for their wholehearted and extremely helpful cooperation and to Professor Tim Maudlin for his helpful advice.

References

(1) N.Liberman,Y.Trope,E.Stephan ,in Social Psychology: A Handbook of Basic Principles, E.T. Higgins, A.W.Kruglanski, Eds.(Guilford, NewYork, 2007), pp. 353–381 ; Y.Trope, N.Liberman, Psychol. Rev. 110, 403 (2003).

(2) Journal of Experimental Psychology: General 1988, VoL 117, No. 4, 371-376, "Phenomenal Characteristics of Memories for Perceived and Imagined Autobiographical Events", Marcia K. Johnson, Aurora G. Suengas, Mary Ann Foley, Carol L. Raye ; and its excellent precursor, still available. Suengras A.G., Johnson M.K, "Effect of Rehearsal of Perceived and Imagined Autobiographical Memories", Paper presented at the annual Meeting of the Eastern Psychological Association, Boston, MA, March 21-24, 1985 ; also a later version of the latter presentation. Suengas, A.G., & Johnson, M.K. (1988). "Qualitative effects of rehearsal on memories for perceived and imagined complex events". Journal of Experimental Psychology: General, 117, 377-389; Johnson's more recent work on statistical learning may also be of interest here. Turk-Browne, N.B., Scholl, B.J., Chun, M.M., & Johnson, M.K. (2009). Neural evidence of statistical learning: Efficient detection of visual regularities without awareness. Journal of Cognitive Neuroscience, 21, 1934-1945.

(3) M.A.Wheeler,D.T.Stuss,E.Tulving, Psychol.Bull. 121, 331(1997); L.K.Fellows,M.J.Farah, Neuropsychologia 43,1214 (2005); D.H.Ingvar, Hum.Neurobiol. 4,127(1985). and details and refs from Gilbert D.T, Wilson T.D., Science (2007), Vol 17, 1351-1355

(4) P.C.W Davies,J.R. Brown, "Superstrings" , 207-8, C.U.P. (1988)

(5) Rashevsky, N., (1961),"Mathematical Principles in Biology and Their Applications", Charles C. Thomas; Rosen R., (1991), "Life Itself", Columbia University Press.

(6) Craig Callender, "Shedding Light on Time", Philosophy of Science, Vol. 67, Supplement. Proceedings of the 1998 Biennial Meetings of the Philosophy of Science Association. Part 11: Symposia Papers. (Sep., 2000), pp. S587-S599.

(7) http://ttjohn.blogspot.com/

(8) Cleeremans A., Haynes (1999) J-D., "Correlating Consciousness: A View from Empirical Science" , Revue Internationale de Philosophie 3 (209):387-420 ; http://srsc.ulb.ac.be/axcWWW/papers/pdf/98-NCC.pdf ; Haynes J-D., (2008), http://medgadget.com/archives/2008/04/not_a_free_will_after_all.html

(9) Frautschi, R.L. Barron's Simplified Approach to Voltaire: Candide. New York: Barron's Educational Series, Inc., 1968 ; "Candide" itself is of course readily available on Project Gutenberg; I could not find a free copy of "Candide" and many other such classics on Google Books.

(10) Shahar Arzy, IstvanMolnar-Szakacs, Olaf Blanke, The Journal of Neuroscience, June 18, 2008, 28(25):6502–6507, "Self in Time: Imagined Self-Location Influences Neural Activity Related to Mental Time Travel "

(11) Metzinger T, (2009), "The Ego Tunnel: The Science of the Mind and the Myth of the Self", Basic Books ; and earlier work.

(12) Noe A., (2009)," Out of Our Heads: Why You Are Not Your Brain, and Other Lessons from the Biology of Consciousness", Hill and Wang.

(13) Rohrer, T., (2006), "The body in space: Embodiment, experientialism and linguistic conceptualization", in "Body, language and mind", Vol. 2, ed. J. Zlatev, T. Ziemke, R. Frank, and R. Dirven. Berlin: Mouton de Gruyter.

(14) Tarnopolsky Y., Grenander U., "History as Points and Lines - Patterns of history and their transformations", (2003), Scribd ; and many recent papers on pattern theory.

(15) Banks W.P., Isham E.A., (2008) "We Infer Rather Than Perceive the Moment We Decided to Act", Psychological Science,Vol 20, Issue 1, Pages 17 - 21 ; Banks W.P., Isham E.A., (2009) "Do we really know what we are doing? Implications of reported time of decision for theories of volition". In: Nadel L., Sinnott-Armstrong W. P.. "Conscious Will and Responsibility: A Tribute to Benjamin Libet". Oxford University Press, in press

(16) Pockett S., Banks W.P., Gallagher S., (2006), "Does Consciousness Cause Behavior?", MIT Press, ISBN: 978-0-262-16237-1

(17) Gilbert D.T., Wilson T.D., (2007), "Prospection: Experiencing the Future", Science, (317), 1351

(18) In Tuszynski's book on consciousness for example, the word 'Penrose' is mentioned some 94 times and the word 'McTaggart' not even once. But Penrose has actually produced no working results on consciousness whatsoever and a study of the work of McTaggart is necessary for all students of consciousness and time; Tuszynski J.A., (Ed.), (2006), "The Emerging Physics of Consciousness" , Springer.

(19) Vohs K.D., Schooler J.W., "The Value of Believing in Free Will : Encouraging a Belief in Determinism Increases Cheating", Psychological Science, (19), 1, p 49 ; commentary at http://www.mindhacks.com/blog/2009/12/the_psychological_ef.html

(20) Yates J., (2009), "A study of attempts at precognition, particularly in dreams, using some of the methods of experimental philosophy", http://philpapers.org/archive/YATASO.1.pdf

(21) Liberman, N. & Trope, Y. (2008), "The psychology of transcending the here and now", Science, 322, 1201-1205.

(22) Yates J., (2009), "The Many Bubble Interpretation, externalism, the extended mind of David Chalmers and Andy Clark, and the work of Alva Noe in connection with Experimental Philosophy and Dreamwork", http://philpapers/org/archive/YATTMB.1.pdf

(23) Norton M.I., Frost J.H., Ariely D., "Less Is More: The Lure of Ambiguity, or Why Familiarity Breeds Contempt", Journal of Personality and Social Psychology, 2007, Vol. 92, No. 1, 97–105

(24) Bargh J.A., "What have we been priming all these years? On the development, mechanisms, and ecology of nonconscious social behavior", Eur. J. Soc. Psychol. 36, 147–168 (2006)

(25) Dhar R., Kim E.Y, "Seeing the Forest or the Trees: Implications of Construal Level Theory for Consumer Choice", Journal of Consumer Psychology, 17(2), 96–100

(26) Liberman N., Trope Y., "Temporal Construal", Psychological Review, 2003, Vol. 110, No. 3, 403–421

(27) Eyal T., Liberman N., Trope Y., "Judging near and distant virtue and

vice", J Exp Soc Psychol. 2008 July 1; 44(4): 1204–1 209

(28) Vallacher, R. R., & Wegner, D. M. (1985). A theory of action identification. Hillsdale, NJ: Lawrence Erlbaum Associates.

(29) Liberman N., Trope, Y., McCrea, S. M., & Sherman, S. J., (2007), "The effect of level of construal on the temporal distance of activity enactment", Journal of Experimental Social Psychology, 43, 143-149.

(30) Liberman N., Trope, Y., McCrea, S. M., & Sherman, S. J., "Construal Level and Procrastination", In press, Psychological Science.

(31) Broemer P., Grabowski A., Gebauer J.E., Ermel O., Diehl M.," How temporal distance from past selves influences self-perception", Eur. J. Soc. Psychol. 38, 697–714 (2008)

(32) Hacker P., Bennett M., (2003), "Philosophical Foundations of Neuroscience", Wiley-Blackwell, ISBN-10: 140510838X, ISBN-13: 978-1405108386

(33) Vul E., Harris C., Winkielman P., Pashler H., (2009). Voodoo Correlations in Social Neuroscience. Perspectives on Psychological Science, in press ; Vul E, Kanwisher N. (in press). "Begging the question: The non-independence error in fMRI data analysis". To appear in Hanson, S. & Bunzl, M (Eds.), Foundations and Philosophy for Neuroimaging.

(34) for example Kircher T., Blümel I., Marjoram D., Lataster T., Krabbendam L., Weber J., Van Os J., Krach S., (2009),"Online mentalising investigated with functional MRI", Neuroscience letters, vol. 454, no3, pp. 176-181

(35) Bass T, (1991), "Newtonian Casino", Penguin, ISBN-10: 0140145931

(36) for example Crutchfield, J. P., Farmer J.D., Packard N.H., Shaw R.S., (1986),"Chaos", Scientific American 255, pp 46-57

(37) "Large Stakes and Big Mistakes" Ariely, D., Gneezy U., Loewenstein G., Mazar M., 'Review of Economic Studies', 76 (2), 2009, 451-469 ; many others such as Ariely D., (2009) "Predictably Irrational", revised edition, Harper, ISBN 978-0-00-725653-2

(38) Woit P.,(2006), "Not even wrong", Basic Books, ISBN-10: 0465092756,

Time Travel in Theory and Practice

ISBN-13: 978-0465092758 ; Peter Woit also maintains a blog of the same title at http://www.math.columbia.edu/~woit/wordpress/ and the title first seems to have been used by Woit in this way in a paper entitled "String Theory, an Evaluation" at http://arxiv.org/abs/physics/0102051.

(39) Woit P., Callender C., (2010), "Philosophy and the String Wars", http://bloggingheads.tv/diavlogs/22449

(40) Arntzenius, F., Maudlin, T., "Time Travel and Modern Physics", The Stanford Encyclopedia of Philosophy (Spring 2010 Edition), Edward N. Zalta (ed.), forthcoming URL = <http://plato.stanford.edu/archives/spr2010/entries/time-travel-phys/>.

(41) www.knowledgerush.com/kr/encyclopedia/Gravitational_singularity/ (Jan, 2010)

(42) http://en.wikipedia.org/wiki/Non-physical_entity (Jan, 2010)

(43) Herrick P., (2000), "The Many Worlds of Logic", p248, Oxford University Press, 2000. ISBN 0-19-515503-3

(44) Maudlin T., (2007), "The Metaphysics within Physics", Oxford University Press, ISBN 978–0–19–921821–9

(45) Orzel C., (2009), "How to teach Physics to your Dog", Simon & Schuster, ISBN-13: 9781416572282.

(46) http://en.wikipedia.org/wiki/The_Fabric_of_Reality ; http://en.wikipedia.org/wiki/Multiverse_(science)

(47) Deutsch D, (1998), "The Fabric of Reality", Penguin, ISBN-10: 0140146903

(48) http://groups.yahoo.com/group/Fabric-of-Reality/

(49) Davies P., (1992), "The Mind of God", p147, Penguin, ISBN 0-14-015815-4

(50) *ibid.*, p152

(51) *ibid.*, p144

(52) Rucker R., "infinity and the Mind", Birkhauser, Boston (1982), p48

(53) Tegmark M., space.mit.edu/home/tegmark/PDF/multiverse_sciam.pdf

(54) Vendral V., (2010), "The Parallel Universe Experiment", http://www.fqxi.org/community/forum/topic/266

(55) Zeh H. D., (2010), "Nonlocality versus nonreality", http://www.fqxi.org/community/forum/topic/323

(56) Maruyama K., Nori F., Vendral V., (2008), "The physics of Maxwell's demon and information", arXiv:0707.3400v2 [physics.hist-ph], 5 Aug 2008

(57) Schmidhuber J., (2010), http://www.idsia.ch/~juergen/

(58) Swarup B., (2010), "The End of the Quantum Road?", http://fqxi.org/community/articles/display/114

(59) Marchal B., (2007), "A Purely Arithmetical, yet Empirically Falsifiable, Interpretation of Plotinus' Theory of Matter", http://iridia.ulb.ac.be/~marchal/publications.html

(60) Vallee J. F., "Incommensurability, Orthodoxy and the Physics of High Strangeness: A 6-layer Model for Anomalous Phenomena", in "Fátima Revisited: The Apparition Phenomenon in Ufology, Psychology, and Science", Compiled by Fernando Fernandes, Joaquim Fernandes & Raul Berenguel, Anomalist Books, ISBN: 1933665238

(61) Vallee J. F., ibid, "No experiment can distinguish between phenomena manifested by visiting interstellar (arbitrarily advanced) ETI and intelligent entities that may exist near Earth within a parallel universe or in different dimensions, or who are (terrestrial) time travelers".

(62) http://golem.ph.utexas.edu/category/

(63) "Free Will", (2010). In Wikipedia, the free encyclopedia. Retrieved February 2, 2010 from http://en.wikipedia.org/wiki/Free_will

(64) Trope Y. , Burnstein E., "A Disposition-Behavior Congruity Model of Perceived Freedom", Journal of Experimental Social Psychology 13, 357-368

(1977)

(65) In Wikipedia, the free encyclopedia. Retrieved Jan 8, 2010, from http://en.wikipedia.org/wiki/Amazon_Mechanical_Turk ; The original "Amazon Mechanical Turk" was devised by Amazon for commercial purposes and can be found on the internet at https://www.mturk.com/mturk/welcome . Further such ideas are at http://polldaddy.com/ , http://www.surveymonkey.com/ , http://lifehacker.com/5451352/become-a-gmail-master-redux? utm_source=feedburner&utm_medium=feed&utm_campaign=Feed %3A+lifehacker%2Ffull+%28Lifehacker%29 and elswhere.

(66) Addis D.R. , Pan L, Vu M. , Laiser N., Schacter D.L., "Constructive episodic simulation of the future and the past: Distinct subsystems of a core brain network mediate imagining and remembering", Neuropsychologia, 47 (2009) 2222–2238 ; and much other work at http://www.psych.auckland.ac.nz/people/donna/donna.htm

(67) Yates J., Patent Number:GB2051465 Publication date:1981-01-14 . I also mention and apply Gott's comment to this patent in http://philpapers.org/archive/YATASO.1.pdf

(68) In Wikipedia, the free encyclopedia. Retrieved Jan 8, 2010, from http://en.wikipedia.org/wiki/Chemical_computer ; Adamatzky A., De Lacy Costello B., Asai T. , Reaction Diffusion Computers, Eslevier, 2005 ; also http://web.mit.edu/newsoffice/2009/ai-overview-1207.html , http://www.physorg.com/news179400180.html and ftp://ftp.cordis.europa.eu/pub/fp7/ict/docs/fet-proactive/chemit-02_en.pdf

(69) Yates J., (2008) , "Towards a Science of Consciousness", p147-8, April 8-12, 2008, Tucson Convention Centre, Tucson, Arizona, Center for Consciousness Studies, University of Arizona. Copy of poster at http://www.scribd.com/doc/2677404/TSC2008 . This also describes the use of the MBI ("Many Bubble Interpretation") to finally resolve the Schrodinger cat paradox.

(70) Ferrari V., Bradley M.M, Codispoti M., Lang P.J.," Detecting Novelty and Significance", Journal of Cognitive Neuroscience, February 2010, Vol. 22, No. 2, Pages 404-411 (doi:10.1162/jocn.2009.21244)

(71) http://dictionary.die.net/pragma suggests that a 'pragma' is a comment

which usually conveys non-essential information, often intended to help the compiler to optimise the program. This usage is only barely metaphorical for SUAC as the 'compiler' there is almost the intelligent human in charge, rather than simply a digital computer ; but as part of a more general argument in the present connection one can see also Frans H. Van Eemeren, Rob Grootendorst, "The Pragma-Dialectical Approach to Fallacies", in "Fallacies: Classical and Contemporary Readings", edited by Hans V. Hansen and Robert C. Pinto (1995).

(72) "Oil -drop experiment", (2010). In Wikipedia, the free encyclopedia. Retrieved February 19, 2010 from http://en.wikipedia.org/wiki/Oil-drop_experiment

(73) Gale R.M., (1968), "The Philosophy of Time, 69 et seq, MacMillan. This is an old reference but covers much work up till 1968. The recent survey by Chalmers (2010), for example, also attempts to imply the traditional theory subdivisions into A-theory and B-theory. My earliest work was based many years ago, partly on communications with the late Arthur Prior, who urged me to work with Robin Gandy, then a Professor of Mathematics at Manchester University, which I did. Arthur Prior might be said to be a 'presentist' and seemed to hold the work of J.N. Findlay in high regard. Indeed in my opinion Findlay's paper [74] provides a nice summary of much important work up to its time of writing.

(74) ibid, p143

(75) Callender C., Edney, F, (2002), especially p66, p109, "Introducing Time", Allen and Unwin, ISBN 1 84046 592 1

Note 1

The title of the present paper ("Not Even Wrong") is of course the same title as that of a recent book by Woit [38] and apparently derives from a catch phrase by Pauli, to the approximate effect that certain theoretical results did not lead to conclusions with respect to substantial and known physical fact. Woit's book is of course about modern string theory and I would add that it is a historical fact that, whilst I was Editor-in-Chief of the "International Journal of Theoretical Physics" for very many years, comments of a somewhat similar nature were often being made to me about string theory (old version). I suppose it is a matter of "O tempora, O mores !" but the world may not have changed much since Cicero's day. Woit's current views on string theory seems to be summarised in an amusing popular video on Bloggingheads [39] by Woit and Callender. Readers may be relieved to find that a detailed knowledge of string theory is not required to read the present peroration, but somewhat less than relieved at finding that the author's view is that, from the viewpoint of enlightened neuroscience, the whole of modern physics suffers the fate that string theory does from Woit's perspective. To start with, modern physics appears have no clear way to even describe let alone deny, prove or delimit human or animal free will and I spend much of the paper dealing with that matter in various ways. The so-called "grandfather paradox", a moderately obvious possible paradox since the days of Godel right up to the modern work of Visser and Thorne, is also confounded, denied and misunderstood in modern physics, even to the point where we are asked to even believe that the denial of even the possibility of human freewill is essential to an understanding of modern physics. I've tried to deal with matters like the "grandfather paradox" briefly in Appendix 2, which in particular discusses the ideas of Maudlin, whose views are actually much more reasonable than those of many other people, though still pretty farfetched.

Note 2

In fact Metzinger in http://socrates.berkeley.edu/~noe/commentaries/NCC-Metzinger.rtf does come surprisingly close to reaching a meaningful compromise, but at the end of the day the required result is completely outside of his frame of reference and he reverts to what amounts to a naive B series or 'flat time' approach.

Note 3

We have a similar sort of objection at the very least. For example, Rucker [52] says "If the Mindscape *(mental space of mathematical objects)* is a One, then it is a member of itself, and thus can only be known through a flash of mystical vision. No rational thought is a member of itself, and so no rational thought can turn the Mindscape into itself". Davies [51] , Tegmark [53] and many others concur to somewhat similar points.

Furthermore, we must bear in mind, as already stated here, that this rather Platonist view of such people as Jeans and Penrose is really a supposition, pure folk psychology and they have no proof of it. It also is becoming a very shaky view and we really do not know of any good proof or justification for the rather complicated arguments needed to bolster it up by now. It could almost be said to be a modern version of epicycle theory.

Note 4

Voltaire's satire is perhaps too bold for today's more thin skinned scientists so I have to refrain from further comment. I have certainly no inclination to emulate the likes of Richard Dawkins or Colin McGinn in satire or disagreement particularly as misunderstanding has made the present topic even more controversial than their own views.Indeed, just as Voltaire [9] lampooned the possible initiator of all our problems - Leibniz - as Mr. Pangloss, so too we today could lampoon the large Roger Penrose style crowd of du Sautoys, and all the rest of them. The name Pangloss translates to English as "all tongue" and "windbag" and one can easily be as exasperated today by such persons - even more exasperated than Richard Dawkins is by Christians and other religious people, since the views of Christians and the like are so different from those of scientists that the difference is usually easy to tell, whereas with neoPanglossians there is the unfortunate fact that they are still leading science astray, as it could be said Leibniz and Newton did, though at least those pair were somewhere near the right course. In effect this motley neoPanglossian rabble comprises most of the mathematically inclined mind theorists today, a few of those who even, like Roger Penrose, seem to be prepared to believe for the sake of their beloved 'B series only mathematics' that conditions in the brain bear resemblance to conditions not too far from absolute zero, and who are quite prepared to cast away freewill so as not to lose even a fraction of the charisma of their precious 'B-series only mathematics'. Like Voltaire we can say "if the applied mindless robot style mathematics of today gives us the best of all possible worlds, then what must the rest of them be like?" The 'many worlds' scientists of today have not been able to avoid this problem either, though I suppose it could be argued that more abstract ideas like complex system theory may have striven to do so, but probably ultimately failed to do so completely. We are truly analog people and have created a digital world of our own. Just as the painter Lowry knew a matchstick world of poverty and starvation, we know a digital world of effective computer science. Lowry at least had the sense to realise that there was also a real world out there and not just his matchstick world of art. In the manner of Thomas Huxley's alleged phraseology where 'Archbishop' is modernised to 'Mathematician', we can say "I would rather be descended from an ape than from a computer". Voltaire says that if people truly wish to think, they should not cling to old and obviously incorrect ideas, but should form their opinions based on experiential knowledge. Voltaire's view should probably also be one of the cornerstones of experimental philosophy.

In very lay terms, modern scientists simply insist that the collection of some scientific data is more real than people's thoughts. It is possibly wrong to insist on this, and a bad early assumption. DT-MRI results, for example, in the way they are produced and calculated, will seem to have some sort of eventual one to one correlation with an updated Leibnizian or Newtonian type of reality. Even Bohm and Einstein, in a way, strove to continually cede first place to Newton or Leibniz and certainly the same is true even with modern string theorists and cosmologists. That fact, whatever the eventual achievements or otherwise of modern string theory and indeed cosmology, has become very true.

Even modern philosophers seem to tend to readily kowtow to this rather notorious idea by effectively putting the heavy horse of philosophy behind the agile and profitable cart of science, whether or not they choose to admit it.

All this is in no way to say that we must put aside the theories of Bohr or Einstein, for example, but we must remember to trim any recognition of their relevance. In a way, for example, the "shut-up-and-calculate" pragma [71] tries to trim that relevance, but clearly leads to puzzlement and to a physics that only relatively primitive entities like a dog can understand [45]. Of course it is important to realise that we are not aiming here to solve philosophical problems created by Bohr's physics - though in a sense we may do so - rather embracing modern physics in its entirety and then, like Oliver Twist, 'asking for more'.

Appendix 1a

31

Experimental Procedure

1. Dream: "Think of a recent dream-any dream you think you can remember fairly well." Fantasy: "Think of a recent fantasy-that is, something you made up and imagined while you were awake-any current fantasy you can remember fairly well."
Unfulfilled intention: "Think of a recent time you intended to do or thought about doing something, but then you never got around to doing it. It should be something you actually might have done but did not." Subjects were asked to remember events like a social occasion, a trip to the library, or a visit to the dentist. Perceived events were selected because they were likely to differ in many ways, for example, degree of social interaction and type and intensity of emotional tone.

2. Subjects were asked to visualise events like a real future social occasion, a real trip to the library, or a real visit to the dentist. Perceived events were selected because they were likely to differ in many ways, for example, degree of social interaction and type and intensity of emotional tone.

3. Subjects were also asked to imagine (at the time of the experiment) the occurrence of events like a dream, a fantasy, or an unfulfilled intention. These imagined events differed in degree of conscious construction and degree of potential realization. Thus we attempted to include a relatively broad representation of events of each type.

Examples of Social occasion: "Think of a recent (or for 2, future) social occasion-party, dinner, or a gathering of some sort that involved more than two people including yourself."
Visit to a library: "Think of a recent (or for 2, future) time you spent in a library."
Trip to dentist: "Think of a recent (or for 2, future) time you visit the dentist."

Now in cases 1 and 2, these are genuine cases which either did happen or will possibly/probably happen, like going to school last week or next week. In fact we include cases like. "Think of when you went to school last week and again last year" and "Think of when you go to school next week and again next year".

Appendix 1b

Experimental Procedure, further details of questioning

Scoring chart

1. This event is 1 = *dim;* 7 = *sharp/clear*
2. This event is 1 = *black and white;* 7 = *entirely color*
3. This event involves visual detail 1 = *little or none;* 7 = *a lot*
4. This event involves sound 1 = *little or none;* 7 = *a lot*
5. This event involves smell 1 = *little or none;* 7 = *a lot*
6. This event involves touch 1 = *little or none;* 7 = *a lot*
7. This event involves taste 1 = *little or none;* 7 = *a lot*
8. Overall vividness is 1 = *vague;* 7 = *very vzvid*
9. The event is 1 = *sketchy;* 7 = *very detailed*
10. Order of events is 1 = *confusing;* 7 = *comprehensible*
I I. Story line is 1 = *simple;* 7 = *complex*
12. Story line is 1 = *bizarre;* 7 = *realistic*
13. The location where the event takes place is 1 = *vague;* 7 = *clear/dzstinct*
14. General setting is 1 = *unfamiliar;* 7 =*familiar*
15. Relative spatial arrangement of objects in my memory for the event is I = *vague;* 7 = *clear/distinct*
16. Relative spatial arrangement of people in event is 1 = *vague;* 7 = *clear/distinct*
17. Where the event takes place is 1 = *vague;* 7 = *clear/distznct*
18. the year is I = *vague;* 7 = *clear/distinct*
19. the season is 1 = *vague;* 7 = *clear/dzstinct*
20. the day is 1 = *vague;* 7 = *clear/distinct*
21. the hour is 1 = *vague;* 7 = *clear/distinct*
22. The event seems 1 = *short;* 7 = *long*
23. The overall tone of the memory is 1 = *negative;* 7 =*positive*
24. In this event I was 1 = *a spectator;* 7 = *a participant*
25. At the time the event seemed like it would have serious implications: 1 = *not at all;* 7 = *definitely*
26. The event does have serious implications: 1 = *not at all;* 7 = *definitely*

27. Any feelings at the time: 1 = *not at all;* 7 = *definitely*

28. Feelings at the time were 1 = *negative;* 7 = *positive*

29. Feelings at the time were 1 = *not intense;* 7 = *very intense*

30. As I am remembering now, my feelings are 1 = *not intense;* 7 = *very intense*

3 1. I remember what I thought at the time: 1 = *not at all;* 7 = *clearly*

32. This memory reveals or says about me: 1 = *not much;* 7 = *a lot*

33. Overall, I remember this event: 1 = *hardly;* 7 = *very well*

34. I remember events relating to this memory that took place: in advance of the event: 1 = *not at all;* 7 = *yes, clearly*

35. after the event: 1 = *not at all;* 7 = *yes, clearly*

36. Do you have any doubts about the accuracy of your memory for this event? 1 = *a great deal of doubt;* 7 = *no doubt whatsoever*

37. Since it happened, I have thought about this event: 1 = *not at all;* 7 = *many times*

38. Since it happened, I have talked about it: 1 = *not at all;* 7 = *marry times*

39. About when did this event happen? Circle one: just today yesterday few days ago last week few weeks ago last

Appendix 2

The "Grandfather Paradox" and similar matters

Now a common word used in physics in such circumstances as the "grandfather paradox" is to say that cases like that are 'unphysical'. Popular encylopedias [41] tend to define 'unphysical' in cases like singularities in general relativity as simply meaning (in the GR case) that general relativity ultimately ceases to be an accurate description of gravity somewhere in the vicinity of what would otherwise be a singularity. Alternatively, encylopedias [42] tend to define "a non-physical entity" as "an entity that lacks a physical or material body or material or physical characteristics. Non-physical entities may be considered hypothetical, e.g. deities of religions no longer conventionally believed in, and used as an example of an imaginary being in analytical philosophy, or they may refer to concepts whose existence is considered in philosophical argument, such as qualia. Or in esotericism they may refer to devas, gods, spirits, and so on, which either lack a body, or possess a subtle body only, and are generally considered belonging to a supra-physical plane of existence. Or in philosophy of mathematics, many people consider numbers,

34

spaces, sets, and so forth to be existent and yet not physical".

Surely all this is simply begging the question or petitio principii, "assuming the initial point", if we intend to try to regard modern physics as a fair description of modern observable phenomena.....

According to Herrick [43] "'seldom is anyone going to simply place the conclusion word-for-word into the premises .... Rather, an arguer might use phraseology that conceals the fact that the conclusion is masquerading as a premise. The conclusion is rephrased to look different and is then placed in the premises".

Maudlin [40] states specifically in his current (2010) update of the Stanford Encylopedia entry on the possibility of time travel that "conceptual and logical "possibility" do not entail possibility in a full-blooded sense. What exactly such a full-blooded sense would be in case of time travel, and whether one could have reason to believe it to obtain, remain to us obscure". So he is not explicit about what, if any restrictions would have to be placed on, for example, general relativity notions, if he cannot find a way round, in real and practical terms, the problems arising from paradoxes like the "grandfather paradox".

But we know that formulations like those of Godel, Thorne or Visser could apparently lead to an unsolved "grandfather paradox" and if we are to believe modern physics, we are left with the fact that, from Maudlin's recent comments at least, they are in no way resolved in it. So modern physics is in fact, as currently formulated, apparently inconsistent and/or incomplete in quite serious ways.

Maudlin's book,"The Metaphysics Within Physics", which basically consists of a ten year collection of some of his essays, has already been reviewed, often very kindly, by many other researchers. But to me, though it is somewhat confusing it must nonetheless be considered, even if there is the feeling that throughout this work that Maudlin may be acting on a very different set of basic premises to myself. I basically have the feeling, which I do not necessarily hold as a philosophy, that the gaps in present day physics are of what one might call a 'Kuhnian' nature, and modern physics and its background of philosophy, psychology and metaphysical what-have-you has gaps which even Thomas Kuhn at his strongest might not have envisaged. For example, even when Maudlin criticises Earman, he does so within a framework of implicit acceptance of a large blob of B series physics, as he seems to make strong arguments invoking general relativity, or at least of some pattern containing general relativity. I will have none of this. The B series is the B series, and acceptable as such up to a point, with its faults, but from the present standpoint we must also consider some version of the A series or at least some partial or restricted mapping or some such thing of the A series.

Later in this discussion I will go into some detail on Maudlin's book where it seems immediately relevant to the matters to hand, but I must pre-empt the comment that my interpretation of the A series is too vague, or that my "Many Bubble Interpretation" is too vague, by pointing out bluntly that B series quantum physics, for example, has had some 80 years to put its house in order since my late colleague Prince Louis de Broglie won his Nobel laureate, and quantum physics has still not succeeded in becoming clear, to the point where the current well-written popular tome [45] "How to teach Physics to your Dog" can actually make quantum physics clearer to a dog than it is to a human being. That does not suggest that dogs have superior insight into modern physics than humans, of course, but it probably shows a lot of things about people that there is probably not time to discuss here, and that fact could be left to further papers on X-phi. To summarise my own views on quantum physics, the fact is that many people still accept the Copenhagen Interpretation, which is an interpretation often known to quantum physics students as the "shut-up-and-calculate" interpretation, and nobody even raises their eyebrows about its popular title any more. Maybe dogs are beginning to acquire more free will than humans have, and humans are dragooning one another into losing their free will. Sartre and the other existentialists of his period may well have thought that to be the present state of affairs.

So quantum physics has had some 80 years to mature to an understandable subject and it has simply got less and less clear. Literally millions of dedicated scientists have had time to clarify it, but problems have got worse and worse to the point of apparently insoluble paradoxes. My position, that we need the A series as well as the B series, has had little work at any time, except by me personally. And I am getting results. I do appreciate the problems of those such as Maudlin with McTaggart's paradox, which is hard to understand and both confused and confusing, and I particularly appreciate Maudlin's points about using the C series rather than the A series, but as a traditionalist I am using the A series as a starter and going on from there. So far I have had success with dreamwork and am in this paper itself carrying out experiments on construal level theory. Further work may relate also to chemical analogue computers, but I see no further quick success so far in that field and a lot more hard work.

Time Travel in Theory and Practice

To return to Maudlin's book,"The Metaphysics Within Physics". The sections of greatest interest are in Chapter 4, on "The Passing of Time". On p109 he says quite blatantly "I believe in a block universe". But this is not enough, nor is a simple 'moving present' if we want to even properly describe the concept of free will - even if some people then propose to disprove or condemn it. Maudlin also, on p109, admits that his views are 'unusual' and and he also says he does not deny the objective flow of time, presumably within some 4d universe.

Now that admission is important, as if time flows, a proper physics should be able to describe that flow. But he seems to totally miss the point that time does not seem to be easily described as one entity, but consists of two different ones, described as A series and B series say. The fact that McTaggart is somewhat muddled, as Maudlin admits, should not obscure for us the fact that time has to be carefully described. Not, as it were, as it was described in the heavy excitement of Gottingen after World War 1, but in the light of centuries of repeated failure by science to quantify freewill and time and the present muddled state of physics in its dealings in particular with quantum mechanics. Perhaps scientists have 'shut-up-and-calculated' for too long, because of impressive results in the short term in large but nonetheless limited areas of physics.

Maudlin's arguments about the passing of time at around p112 seem rather vague, along the lines of a philosopher having to cope with nonexistent physics. I would thus take the view that the problem is not with Maudlin, but with physics, and that I am correct in trying to put it right by using the A series in addition to the B series. Perhaps as somewhat of a sop to Maudlin, I would concur that this may not be the only way to solve this problem ! But here I bow to McTaggart, and suggest his thinking, perhaps somewhat crude by today's standards, can be taken as being along the right lines - by following his leads, the existence of freewill is maintained and the work is in producing more physics, which I am showing can be done and is indeed successful in producing results to date [7,20,22], and in the present paper.

On pp 158-8 Maudlin goes on to say "all God did was to fix the physical laws and the initial physical state of the universe, and the rest of the state of the universe has evolved (either deterministically or stochastically) from that". Now we are thus clearly left with a denial of free will in Maudlin's theories. That is somewhat of a shame, as he seems to be renouncing fact for mathematical fiction, even if in practice he seems to try to post enough philosophical provisos so that he can change his mind later if he has to.

The Deutsch 'multiple universe' theory in fact seems at the moment the most likely of many other 'multiverse' approaches, most of which make use of multiple artificially mathematically created universe without any real known substance of any sort, but clearly the 'block universe' type model leaves such systems or worldviews clearly in the realm of B series physics and so my usual objections still apply in general.

Two wikis [46] and a book [47] sum up the Deutsch "multiverse hypothesis" and even by implication and reference other approaches such as that of Rees or Tegmark, so I will not recapitulate the Deutsch approach. It's pretty hypothetical and the term 'Occam's razor' is frequently used in connection with it for obvious reasons. However the "grandfather paradox" does not seem to apply to the Deutsch multiverse, and this fact relates to the enormous number of postulated worlds. The existence in actual fact of these postulated worlds is a matter of some debate within the theory. And then there are, if you want there to be, the looming dilemmas of all the philosophy attached to such a matter. Maudlin's work cited above [44] is one way such matters can be handled. There's a discussion group [48] about the Deutsch multiverse called "Fabric of Reality" which I read and often contributed to for many years.

Vlatko Vendral [54] and others try to consider an interesting test for the existence of multiverses of one kind and another. Such results have been considered seriously by those such as Dieter Zeh [55]. Vendral's essay [56] on Maxwell's demons is certainly interesting if viewed in the light of modern quantum computing. Jurgen Schmidhuber [57], Caslav Brukner [58], Bruno Marchal [59], Jacques Vallee [60] and many others have also tried to follow similarly difficult paths and of course much positive comment could be made about their work. Vendral [54], however, suggests that perhaps some of such authors are leaving the mainstream of science for speculation, which is harmless but normally insufficiently rewarding. Arthur C. Clarke, H.G. Wells and other similar writers may well have inspired further research and even have been interesting to scholars, but statements like that of Vallee [61] seem unlikely to become of important use for further immediate research, but are closer to futurology, film scripting or simply science fantasy. To be overambitious can be unhelpful to immediate progress and to seek hostages against posterity in this way can easily overshadow real merit. All this seemingly sends working scientists up a metaphorical Tower of Babel, rather than allowing them, like stout Cortez allegedly did, to "stare with wild

38

surmise" at something new to them, like the Pacific Ocean would have been to Cortez. Or, indeed, to do as Cantor did, and some would say is being done even now in the n-category cafe [(62)]. However, none of our current investigators in the area, other than myself, seem to even try to use the A series. One reason that their results are so far simply rather hypothetical may well be that they have not dealt with the A series nor even considered people's human characteristics adequately. Instead they tend to grope for ideas like 'human freewill' and 'god' within the rather stultified arena of existing mathematical formalism - or try to create new, and even more stultifying formalism. Cantor and to a lesser extent Godel and Chaitin have shown that the power of mathematics can bring us to a fresh arena of thought, rather than to a metaphorical Tower of Babel as Woit might well claim has now happened with string theory, for example. Towers of Babel may indeed be fascinating, especially to mathematicians, but are off topic here. Such speculation is not to be condemned, of course, but it can be rather a pity.

Appendix 3

Professor Maudlin's comments

There is clearly much too much to comment on, but let me make two brief points:

1: McTaggart's confusion is easy to state: he write as though there is exactly one "A-series", which changes through time. This is incorrect. The A-series consists in the events in the history of the universe categorized as (in the simplest case) "past", "present" and "future". So even from the perspective the physics McTaggart knew (Relativity is not very important here), there are an infinitude of different A-series: on for each moment of time categorized as "present". In sum, given the B-series and a single event (or single moment) to count as "present" you can define an A-series, and since there is an infinitude of such events, there is an infinitude of A-series. But each A-series postulates nothing more in reality than one already has in the B-series. As I mentioned, the B-series has an intrinsic direction- the basic asymmteric relation of "earlier than". So "starting with the A-series", you have no more to work with than someone who starts with the B-series and has tken-reflective terms like "now".

2. As a compatibilist, I do not think any issues about time or determinism have any bearing on the issue of free will. Indeed, I do not think there is any such issue. Nothing in physics prevents the description of humans as deliberating

about different courses of action, evaluating the foreseen outcomes, or possible outcomes, and acting on that evaluation. As Hume points out, this just is free will, which we have if we are not a prisoner in chains, and are capable of this sort of deliberation.

Reply to Professor Maudlin

1. In effect Professor Maudlin seems to be saying that he thinks that anything useful in the A series can also be conveyed in the B series. I have no argument with that in principle, though it is conjectural until specific cases are established. But equally, we might have well been using pre-Kepler epicycles nowadays to describe planetary motion and might well ultimately have very similar physics to what we have now !

In mental terms, though, the crux of it is that the A series and the B series 'look' different and the physical techniques of modern mathematics seem to describe fairly well the physical movements of physically apparent entities like planets etc. but are shrouded in mystery when it comes to dealing with the human mind.

In my opinion the experimental philosopher must adopt the stand of considering things as they are and not simply on the abstract plane of thought - metaphorically his armchair must burn, to adopt a current phraseology - and he must deal directly with the mind.

Obviously I have done my best to provide a B series mapping of A series concepts, using my Berkeley Madonna models and other models, but there is a long history of cases of persons who are, or claim to be, A-series supporters, B-series supporters, and A and B series supporters. It is clear that believed differences in A and B series have taxed the minds of these, often prominent people. The book of Gale [73] provides many examples of this fact, and Professor Maudlin might well refer rightly to some of these as being due to 'confusion'. The matter is somewhat simplified in the approach in Callender's elementary introduction [75] to time where in common with many, Callender tries to speak of 'tensed' and 'tenseless' theories of time as if these have real differences and each have possibly valuable properties of their own. Then, maybe, we are given the thought fom Callender [75] that one view or the other is superfluous or of less basic merit. Like Professor Maudlin I want to resolve the problems, but am not, at this stage, prepared to cut the Gordian knot but rather prefer to examine the real implications of the ideas associated with the A series as far as mental awareness and understanding are required.

I really have to mention Putnam's comments on 'tensed' time which arise through special relativity and perhaps in other ways, as they inspire interest and suggest further examination rather than simple dismissal of the A series. In fact, rather than suggesting the abandonment of the A series, they suggest that important differences may exist between A series and B series and may even enforce the idea that the B series describes a simplistically devised "physical" world and the A series is more in concordance with a "mental" world of consciousness but a detailed program would need much thought and a simple approximation might only too easily lapse into naivety.

I believe that the difference or otherwise between A and B series may be one of today's cardinal problems in experimental philosophy, and for once some sort of solution, at least of a currently expedient nature, may be obtainable and even inherent in my present studies. I hope that this is not too weak a statement as I expect actual physical results and appear to be obtaining these gradually.

2. On compatibilism, Professor Maudlin's view seems to be probably a healthy one and ultimately perhaps correct. Indeed, there probably may not be a problem with time, provided we can consider it in the right way. But this may need the A series as well as the B series, at least to be going on with.

2.02 Simple Robotic Simulation of Perception, and Time Travel

Abstract

Simple NXT robots in configurations based on a type 2 Braitenberg configuration are used to simulate problems in time travel, flash-lag effects and other situations. The philosophical implications are briefly explored.

Introductory Comments

As Dawson (2010) shows, qualities such as free will, discernment, moderation, self-recognition and narcissism can apparently be created in a robot using 40 lines of code, two motors and two sensors. DuPuis (Dawson, 2010a) gives a simple explanation in a very short video.

There seems no reason why much more complex ideas, such as physically travelling backwards and forwards in time, cannot be incorporated into such a system as required. On a less ambitious basis, mental time travel, both to the future and to the past, seems as if it may have somewhat similar psychological parameters associated with it (Yates, 2010). This process is almost the reverse of experimental philosophy. One alters or indeed creates a simple system to obtain just a few ideas which a philosopher could relate, if he wished, to properties of a mind. DuPuis (and for that matter Dennett (1993) and Braitenberg) refers to similar ideas as "The law of uphill analysis and downhill synthesis".

Continually one reads in papers on very simple robots a few lines to the effect that 'no memory has been incorporated into the system but it almost appears to have a memory of its own' (e.g. Wikipedia, 2011), Nowadays people frequently even refer to embodied cognition, i.e. intelligent behavior that emerges from sensorimotor interaction between the agent and its environment, without any need for an internal memory, representation of the environment, or inference (Sutton, 2006).

Clearly we can abuse this facility by holding the robot in suspense of its actual circumstances and looking for a reaction - to make it try to 'foresee the future', for example. or for other psychological experiments.

Further, as Fodor (1997) points out "not all the functions of a classical computer can be encoded in the form of an explicit program—some of them must be wired in". Many others including myself (Yates (2010) as stated in "Not Even Wrong - a view of current science of the mind") and others like Andy Clark and David Chalmers, would probably see this as at least a possible and reasonable viewpoint - and indeed would frequently go much further and more widely. I certainly would.

I deal with General Theoretical Concerns in Appendix 1 and Appendix 4 of this paper, the Flash-Lag Effect in Appendix 2 and A Relationship to Dreams in Appendix 3.

Experimentation

Many people have used Lego robots for scientific experimentation by now. Stafford (2010) points out that there can be wild swings when slightly different robots are used, which are duplicates of one another, and gives a graph which shows the wild swings he has obtained. We find that even with the same robot, and mildly varied parameters and circumstances, swings can also be obtained. This is unsurprising, given chaos theory and the many other factors we have discussed in earlier papers.

Stafford also points out that experimental results with real people are likely to be even more varied. All this is normal enough in the field of experimental psychology, but the immediate result is that staying with simple Lego robots may make ultimate results slightly more significant. Dynamic systems theory indicated already that the use of further elaborate mathematical structures is unlikely to help immediately in most cases (Hannon , 1997). In short to build a simulated robot using simple computer mathematics probably will not help at this stage although the possibility must be retained.

The flash-lag effect was simulated in several ways. But to begin with it was done by allowing the NXT robot to experience colored lines and marks on the ground, using a color sensor triggered by viewing close proximity to the red lines or marks. This caused the robot to jump backwards (effectively immediately) but to retain the memory of previous stimulations to the right and left light sensors, which – other than the repulse reaction – were postdated for times from 80 milliseconds to 30 seconds. The Lego robot used in the early stages was an ordinary Braitenberg configuration, approximating to a type 2 Braitenberg vehicle. . We were prepared to use a Bluetooth connection to a computer running SciLab , but for the present simple purposes enough information could be programmed directly into the NXT brick. So in effect the robot's internal time was dated 80 milliseconds to 30 seconds earlier than the 'surrounding' time, except when the red ground was contacted. So we simulated a flash-lag effect for a reaction time of 80 milliseconds to 30 seconds. We can even simulate various forms of time travel using much the same equipment.

The possibility of adding 'stochastic resonance' effects to sensitise interactions, the use of various optical optical illusions, 'gravity hills', and even the Shams 'ventriloquist illusion' (Wozny, 2010) can be, and was, replicated on these very simple models. Most authors seem to have simply elaborated details of the Braitenberg vehicle, somewhat after the manner that Braitenberg himself tried to describe more and more complex versions of his vehicle, but an alternative approach is to simply alter the environment or the type of interaction with the environment without adding undue complexity to the model.

Most of these robots do appear spontaneous and often the robot could seem – somewhat anthropomorphically - to be trying to work out how to escape from its enclosure and seems to be persistently looking for various alternative ways to do so !

The present experiments are using a wireless electronic marker to allow precise graphical plots so that a family of curves becomes available, for all types of mathematical analysis.

Acknowledgments

I would like to thank Tom Stafford for his helpful advice and comments on this work.

References

Changizi, Mark A. et al. (2008), "Perceiving the Present a Systematization of Illusions". Cognitive Science 32,3 : 459-503.

Clayton, N. S., Bussey, T. J., & Dickinson, A. (2003). Can animals recall the past and plan for the future? , Nature Reviews Neuroscience, 4(8), 685-691.

Dawson M. , Dupuis B., (2010) "From Bricks to Brains" Chapter 6, 'Grey Walter Tortoise' , "Tortoise Behaviour", http://www.bcp.psych.ualberta.ca/~mike/BricksToBrains/Video/Chap7/Video7-2.mpg

Dawson M. , Dupuis B., (2010a), "Robots call the Shots", http://www.youtube.com/watch?v=Nxbd3jGCSFY

Dennett D.C., (1993), "Consciousness Explained", p171, Penguin Books, London, ISBN 0-14-012867-0

Eagleman, D. M., & Sejnowski, T. J. (2000a), "Motion integration and postdiction in visual awareness", Science, 287, 2036–2038.

Eagleman, D. M., & Sejnowski, T. J. (2000b), "The position of moving objects", Science, 289, 1107a–1107b.

Eagleman, D. M., & Sejnowski, T. J. (2000c), "Flash-lag effect: Differential latency, not postdiction", Science, 290, 1051a–1051b.

Fodor J. (1997), in Haugeland J., "Mind Design II", p342 , Bradford Books, MIT Press.

Giere R.N., (2006), "Scientific Perspectivism", University of Chicago Press, ISBN 0226292126

Hannon B., Ruth M., (1997) , " Modelling Dynamical Systems", p48 et seq, Springer, New York

Honderich T., (2011), "The Determinism and Freedom Philosophy Website", http://www.ucl.ac.uk/~uctytho/dfwIntroIndex.htm

Kühn, S., & Brass, M. (2009), "Retrospective construction of the judgement of free choice", Consciousness and Cognition, 18, 12-21.

Marsh A. A. , Kozak M.N., Wegner D.M., Reid M.E., Yu H.H., Blair R.J.R., (2010), "The neural substrates of action identification", Social Cognitive and Affective Neuroscience, February 11, 2010

Martin-Ordas, G., Haun, D., Colmenares, F., & Call, J. (2010), "Keeping track of time: evidence for episodic-like memory in great apes", Anim Cogn, 13, 331–340.

Nijhawan R., (2008), "Visual prediction: Psychophysics and neurophysiology of compensation for time delays", Behavioral and Brain Sciences 31, 179–239, doi: 10.1017/S0140525X08003804

Nyberg L, Kim AS, Habib R, Levine B, & Tulving E (2010). Consciousness of subjective time in the brain. Proceedings of the National Academy of Sciences of the United States of America PMID: 21135219

Pockett S., (2002), "Backward Referral, Flash-Lags, and Quantum Free Will: A Response to Commentaries on Articles by Pockett, Klein, Gomes, and Trevena and Miller", Consciousness and Cognition 11, 314–325 , doi:10.1006/ccog.2002.0562 ; Pockett S., Banks W.P., Gallagher S., (2006), "Does Consciousness Cause Behavior?", MIT Press, ISBN: 978-0-262-16237-1

Ramesan (2010), "Yoga-based and Knowledge-based spiritual paths", http://beyond-advaita.blogspot.com/2010/12/yoga-based-and-knowledge-based.html ; and other blog items.

Stafford T., (2010) , "Mindhacks" , http://mindhacks.com/2010/06/23/the-scientific-method-lego-robots-edition/

Suddendorf T., Corballis M.C.,(2008), "Forum : New evidence for animal foresight?", Animal Behaviour, 75, e1-e3

Sutton J., (2006),"Introduction: Memory, Embodied Cognition, and the Extended Mind", Philosophical Psychology Vol. 19, No. 3, June 2006, pp. 281–289

Szpunar K.K, (2010), Memory & Cognition, 38, 531-540.

Vallacher, R. R., & Wegner, D. M. (1987), "What do people think they're doing? Action identification and human behavior". Psychological Review, 94, 3-15.

WikiBooks, (2010), "Consciousness Studies/Neuroscience 2" , http://en.wikibooks.org/wiki/Consciousness_Studies/Neuroscience_2 , 25 December

Wikipedia, (2011), The Free Encyclopedia, Wikimedia Foundation , , March 29, 2011

Wozny D, Beierholm U, Shams L., (2010), "Probability matching as a computational strategy used in perception", PLoS Comput Biol 2010;6(8):e1000871. doi:10.1371/journal.pcbi.1000871.

Yates, J. (2008), "Category theory applied to a radically new but logically essential description of time and space", http://cogprints.org/6176/ , PHILICA.COM, Article number 135

Yates J., (2010), "Not Even Wrong - a view of current science of the mind", http://philpapers.org/archive/YATNEW and on website http://www.ifsgoa.com/publications

Yates J., (2010a), "The Andromeda Paradox, Bricolage, and Perspectival

Realism", http://www.ifsgoa.com/publications

Appendix 1

General Theoretical Concerns

Szpunar (2010) and many others points out that the contents of memory appear to be routinely sampled during the construction of personal future scenarios. There are many more papers (by Donna Addis, Chris Frith, etc) which attempt to delineate further details of such an idea and some of these may be useful for the construction of further robots.

Many people (Changizi, 2008) claim that humans already have a slightly developed sense of 'precognition' which helps them to survive, and indeed some people (e.g. Suddendorf, 2008) claim that this is also true of animals.

Whilst the amount of time this 'precognition' occurs in humans is quite small, being in the region of 80msec there is no reason why the effect cannot be incorporated into a robot for much longer periods. Folk psychology would indicate that the 80msec lag is simply a factor dependent on internal brain design but that seems to be an assumption based on B series Newtonian physics results. We already know that special relativity limits the use of Newtonian physics and counterfactual results may be obtained , e.g. in the case of the Andromeda paradox, if even special relativity is accepted as having complete rigor in evidently unsuitable circumstances.

The present work also bears in mind the work of Nyberg (2010) who examines (using fMRI) brain regions which seem to be similar to those we described in an earlier paper using experimental philosophy. But his work does not seem to show what the area is actually doing, which could need the use of appropriate brain damaged patients or patients with temporary brain damage, as could perhaps be imposed with rTMS.

In actual fact Nyberg's results showed that the left lateral parietal cortex was differentially activated by nonpresent subjective times compared with the present (past and future > present). A similar pattern was observed in the left frontal cortex, cerebellum, and thalamus. There was no evidence that the hippocampal region is involved in subjective time travel. These findings provide support for theoretical ideas concerning chronesthesia and mental time travel

Further studies on this matter are often carried out using yoga and other meditational techniques, often considered as fringe science susceptible to mainline studies such as rTMS and fMRI. (Ramesan, 2010).

Here we use Occam's razor and try to avoid extra Newtonian assumptions about our system.

Appendix 2

The flash-lag effect as a typical case

The flash-lag effect has been interpreted in many ways (Pockett, 2002). To summarise:
1. One hypothesis is that there are differences in the visual persistence of flashed and moving stimuli.
2. A second hypothesis is that subjects extrapolate the motion of a moving stimulus about 80 ms into the future.
3. A third hypothesis is that the visual system takes longer to process a flash than a moving object.
4. A fourth hypothesis, called by its authors the postdiction hypothesis (Eagleman & Sejnowski, 2000a, 2000b, 2000c), is that at any particular point in time the percept of a moving object involves a time-weighted integration of all the information about that object which became available during the previous 80 ms.

Further, Changizi (2008) even asserts that the human visual system has evolved to compensate for neural delays, generating images of what will occur one-tenth of a second into the future. This foresight enables human to react to events in the present.

Nijhawan (2008) gives a large number of references in his review of the psychophysics of neural prediction and even goes so far as to say "Visual prediction has a strong logical basis and seems consonant with other visual phenomena ....Prediction may be a multi-level, multi-modal phenomenon found in both sensory and motor systems. ... This general approach to the study of prediction suggests possibilities that could unify research from single cells to cognition".

Setting up a case using the Lego models may allow the expansion of the 80ms time scale to a very large figure, and simulate the further development of machine or human intelligence.

Clearly plenty of normal time-series prediction methods can be incorporated in any such program. We use open-source SciLab rather than MathLab for such work but generally either will probably do. In fact in this instance we simply used Lego robots with time delays from 1msec up to 100 seconds. These delays can often readily be incorporated into the robot itself without the need for a Bluetooth connection to a computer.

Appendix 3

Theoretical Concerns regarding Dreams

The literature contains many references to the effect of external stimuli upon dreams, particularly REM dreams. Anecdotal information suggests that external stimuli may relate to dreams which occur roughly at the same time as the external stimulus. However the kind of dreams which seem to relate to this effect seems (e.g Solms) to be relatively uncommon and the idea that the dream occurs before the stimulation is as yet unproven.

The order of pre-awakening stimulus and dream seem to have been most frequently found using dream laboratories to be: First-stimulus. Second-relevant dream. The jury is of course still out, regarding the overall picture.

However the dream as anecdotally described so often seems to contain a long scenario. So there is still the mystery as to how the mind sets up such a scenario so fast. But this is the sort of problem we are always encountering during studies of consciousness. For example WikiBooks (2010) points out: "When a dog barks we see its jaws open at the same time as the bark and both jaws and bark are at the same location. We take this for granted but the brain must be engaging in some complex processing to achieve this synchronised and appropriately positioned set of objects and events."

Now the apparently unexpectedly fast degree of human brain comprehension could be due to strange microtubule effects, or even quantum computational effects, but it seems more likely that it is due to an inadequate formulation of the McTaggart A series. Just as Newtonian methods were not good enough to explain special relativity, so too may contemporary physics be - as yet and possibly always - incapable of adequately formulating the A series.

We have already mentioned and will discuss the apparently high levels of anthropomorphism and indeed apparent innate philosophical ability encountered with Braitenberg type 2 robots so an obvious step will be to use

BVs to attempt to simulate various types of perception, including anomalous perception, to see what we get.

For a start, from the 'flash-lag effect' and the 'crank handle effect' it is obvious that strange delays in perceived results appear to relate to suppressed or unorganised perceptual results. An obvious approach is to obtain visual results but delay them relatively to the knock-on or immediate results of contact. We may even be able to simulate dream perception using BVs using variable delay times.

It must be stressed that the use of digital simulations is NOT sufficient for these experiments and we need real machines as I pointed out to begin with.

Appendix 4

Further justifications of using NXT brick and similar devices

The complex philosophical problems involving subjects like free will, the subconscious (if any) and action identification can only be mentioned in this brief work without detailed discussion or commentary. On the general subject of free will, the basic approach of Peter Strawson and more advanced models such as the attempts of Kane, Wegner , Marsh (2010) , Kuhn and Brass (2009) could be dealt with in detail as could the William James basic model which was of course refined by such as Margenau, Dennett, Popper and Heisenberg. Then there are illusionists like Smilansky. And as a sort of summing up there is of course Honderich's (2011) website – which promotes hard determinism, though somewhat leniently.

Action identification means that the actor is always sensitive to contextual cues at higher levels of identification but moves to lower levels of identification if the action proves difficult to maintain with higher level identities in mind.

Thus it is interesting to see what would happen at very basic levels of programming of the NXT brick, if we assume some merit for Wegner's (1987) idea.

To show how perspectivalism (or what Giere calls perspectivism) works, Giere (2006) suggests that colours , for example, are really the interaction of the world and the human physical system. He then generalises the argument to scientific observation and suggests for example that the output of scientific instruments is perspectival. He reckons that models based on complex scientific theories like that of Maxwell can be used to devise models which can make various claims about aspects of the world.

In an earlier paper we described the merits and actual value (in that we resolved the Andromeda paradox) of an approach like that of Giere (Yates, 2010a).

Thus we base a potential model, not simply on modern physics which is manifestly quite different to neurological observations although these are frequently explained in terms of modern physics, but on simple real models like the NXT brick, set up perhaps as a type 2 Braitenberg vehicle or something of similar general nature.

The work of Clayton (2003), Martin-Ordas (2010) and many others, suggests that the so-called 'mental time travel' of Tulving can occur for animals, 3-5 year old children and even birds. If so, a simpler model should perhaps also work for the NXT brick, after a fashion as there is no immediate imperative of some special additional concept like 'qualia', whether present or not. We must beware not to revolt from this idea, remembering the 'uncanny valley' effect. Certainly the possibility seems easier to understand than the somewhat muddled ideas of determinism, freewill, and the lack of either which abound everywhere. Also it should be testable with some ingenuity and mild difficulty. In earlier work (Yates, 2008) we already gave a mathematical attempt to describe the so-called 'unconscious' mind and its relationship to the conscious mind using complex system theory and a model based on potential reactions between a 'Romeo' and a 'Juliet' model. Here there seems to be no reason why a roughly mathematical model – and hence possibly a real physical model – should not allocate 'conscious' and 'unconscious' elements to a simple NXT robot.

2.03 The Andromeda Paradox, Bricolage, and

Perspectival Realism

Abstract
Previous failures to resolve the Andromeda paradox are discussed, and a resolution is presented using bricolage.

Introduction and Modus Operandi

Paul Nahin (1998) gives an amusing historical account of the Rietdijk/Putnam paradox, with many references. Nahin portrays it somewhat as a comedy of philosophical errors, involving tolerably well known names like Putnam, Rietdijk, Harris, Capek, Sklar, Stein, Fitzgerald, Weingard, Earman, Godfrey Smith, and Smart. But much mirth has to be cautioned as we must bear in mind that this is simply normal and very reputable progress in the field of scientific discovery.

The Rietdijk/Putnam paradox has become more latterly known as the Andromeda Paradox or the Rietdijk/Putnam/Penrose paradox and of course there are still 'philosophical' problems. A current reasoned description is that of Savitt (2006), which by and large defines the current situation at the time of writing. Bresnard (2010) and others continue to ponder the matter.

Huw Price has considered causal perspectivalism as a way to resolve apparent discrepancies in physics, particularly in quantum theory. Price (2005) has written enough to establish that there is a case for using perspectivalism in one form or another to deal with such cases.

Without wishing to go over the merits of various possibilities of perspectivalism we are immediately led to the work of Caspar Hare (2010). His notion of 'eternalist A-theory' (as distinct from 'eliminativist A-theory' or 'presentism') would seem to be almost part of a description of our own MBI (Many Bubble Interpretation), in the sense that the MBI can be taken as implying a form of his 'perspectival realism', discussed in Hare's simply written paper.

It is important to be clear that this does not infer that we support some form of "murky relativism" (Hackenberg, 2010) for which Hackenberg's example of an extreme form would be to grant equal time to all areas of human knowledge, such as astrology and astronomy.

52

An example of Giere's (1999) view of perspectival realism is that vision of the color of an apple, for example, is a psychological property of the interaction between the apple and the observer and that it does not exist apart from either. Returning to the Andromeda paradox and Hare's (2010) paper we clearly take the option suggested by Hare that special relativity does not tell us all we need to know about simultaneity, thus avoiding any direct conflict with the Andromeda paradox whatsoever. So in the MBI there may be no problem with the Andromeda paradox. There certainly does not seem to be any problem. As Hare points out, denial that special relativity is a complete description is an option that can be taken by Hare's 'tense realist' as well, as it happens. Of course tense realism is in some ways quite close to our position of perspectival realism which refers also to things of which one is not perceptually aware, such as, indeed, a possible future invasion by Andromeda to use the 'Andromeda paradox' example.

As Petkov (2009) has made clear, there is a lot that special relativity does not give to us that we may really feel we should know but that special relativity will never vouchsafe to us, the actual value of the one-way velocity of light being just one example. This situation may be felt unsatisfactory though any dissatisfaction could perhaps be written off by some philosophers as a matter of 'folk psychology'

Other Approaches

Zimmerman (2007, 2010) perhaps presents the other most current detailed attempt to further resolve these problems. Other approaches such as that of Craig (2001) include the use of Lorentzian space-time and seem to carry as many or more problems.

Zimmerman (2007) says " I do not believe the A-theory automatically requires a return to Lorentz; and I try to explain why elsewhere. Granted, the A-theorist attributes a special status to one way of slicing the manifold. But this structure can be added without thereby undermining relativity's account of the way space-time works; the causal role assigned to space-time by relativity is consistent with a privileged slicing. The A-theorist's additional fundamental structure can, in principle, leave the web of relativistic spacetime distance relations intact – still doing its intended job in explanations of why things move in the ways they do".

Zimmerman (2010) also adds "If, as A-theorists believe, there is an objective fact about what is presently happening, there must be an objective fact about which events are simultaneous with one another — in other words, a fact about simultaneity that is not relative to anything, including the frames of reference of SR, or the local frames of GR. But, on the face of it, these scientific theories require that simultaneity be frame-relative."

Zimmerman tries to include his 'A-theory type' views in the special/general relativity approach by affirming "There is an objective, important difference between events that are really happening to me, and ones that merely did or will happen to me; and the events that are really happening to me are confined to a tiny region, r, on the world-line I will eventually have traced through the manifold."

He can of course do so, but the principle of parsimony suggests that for the moment it may be simpler to consider a more general approach as we do here, and at least for the moment take the option that special relativity does not tell us all we need to know about simultaneity.

Bricolage

Bricolage is the art of tinkering (or improving, or improvising) with what is to hand to improve it.

On bricolage, Deleuze (1983) quotes Levi-Strauss where bricolage is defined as almost just a feature of the home workshop, and one feels there is some element almost of both disdain and mild admiration, a view which also seems to be focused on by Kauffman (1995), though Dennett rather disapproves and refers to Kauffman as a "meta-engineer". Elsewhere, however, Dennett (1998) even uses the word 'kludge' rather disrespectfully but still offers obeisance to Gould as an early describer of Nature's bricolage. Dennett seems to realise that philosophers have somewhat of a blind spot to bricolage, and indeed even refers to Putnam in that connection !

Thus there has clearly been an unfortunate tendency for philosophers and natural scientists to depersonalise physical experience and importantly Giere and other perspectival realists have tried to temper and usefully modify that tendency. This would seem to be a genuine psychological problem, of a somewhat similar type to that which Copernicus found with the Catholic Church.

Derrida seems to begin to see a rift where he says (in "Structure Sign and Play in the Discourse of the Human Sciences") "there are two approaches to inquiry which are absolutely irreconcilable, that which seeks the origin or center, and that which affirms play and becoming". Derrida actually suggests that all discourses are bricoleur, a view to some extent mirrored in Varela's (1995) own view that "Organisms have to be understood as a mesh of virtual selves. I don't have one identity, I have a bricolage of various identities. I have a cellular identity, I have an immune identity, I have a cognitive identity, I have various identities that manifest in different modes of interaction. These are my various selves. I'm interested in gaining further insight into how to clarify this notion of transition from the local to the global, and how these various selves come together and apart in the evolutionary dance"

So in further experiments we have to tinker with various mental process in different ways, perhaps directly using some of the methods of experimental philosophy in questionnaires and other ways (Yates 2008, 2009, 2010 etc) but also by other methods such as robotic simulation and tinkering, techniques summarised to an extent in the work of Chalmers, Clark and of course the important work of Dawson (2010). In short there is not much use in hoping that special relativity alone would revise the processes of the mind. It is essential to beware of the 'uncanny valley' phenomena often written large for philosophers and even natural scientists, and to dare to carry out effective robotic mind simulation techniques without depersonalisation worries or fears, though as the old saw goes 'the price of freedom is eternal vigilance' rather than ignorance, superstition or untoward enthusiasm. So in our new laboratories at Chandor, Goa, we are using robotic techniques after the manner of Dawson to begin with.

Conclusion

Our MBI interpretation resolves the Andromeda paradox, just as it was able to explain the Schrodinger Cat mystery (Yates 2008, 2009, 2010). There seems remains a problem that, in the way of 'resolution' of such paradoxes, there is still the feeling that there 'was something there to look at'. We have carefully defined what that 'something' is - basically a misplacement of worries about special relativity. We have also defined at least one way of dealing with that 'something' - the way being bricolage. We proceed further with this way in a later paper in this series.

Acknowledgments

Time Travel in Theory and Practice

I would like to thank Professor Dean Zimmerman for his helpful advice and comments on this work.

References

Besnard F., (2010), "Time of Philosophers, Time of Physicists, Time of Mathematicians", preprint, and slide version at Frontiers of Fundamental Physics, 11th Symposium

Craig W. L., (2001), "Time and the Metaphysics of Relativity", Dordrecht: Kluwer Academic Publishers

Dawson, M.R.W., Dupuis, B., & Wilson, M. (2010). "From Bricks to Brains: The Embodied Cognitive Science of LEGO Robots", Edmonton, AB: Athabasca University Press

Deleuze G., Buattari F., (1983), "Anti-Oedipus: Capitalism and Schizophrenia", p7, University of Minnesota

Dennett D., (1998), "Brain Children", p269, Bradford, MIT Press, Cambridge, Massachusetts

Giere R.N.,(2000), "The Perspectival Nature of Scientific Observation", Annual Meeting of the Philosophy of Science Association, Vancouver, British Columbia, November 2-4, 2000.

Hackenberg T.D., (2009), "Realism Without Truth: A Review of Giere's 'Science without Laws' and 'Scientific Perspectivism'", Journal of the Experimental Analysis of Behavior, 91, (3), 391–402

Hare C., (2010), "Realism about Tense and Perspective", Philosophy Compass, March 2010

Kauffman S., (1995), "The Emergent Self", chapter 20, excerpted from Brockman J., "The Third Culture: Beyond the Scientific Revolution", Simon & Schuster

Nahin P., (1998), "Time Travel In Physics, Metaphysics And Science Fiction", p 173 onwards and elsewhere, American Institute of Physics.

Petkov V., (2009), "Conventionality of Simultaneity and Reality", philsci-archive.pitt.edu/3986/1/elsevier2.pdf

Price H., (2005), "Causal Perspectivalism", in Price H., Corry R., eds., (2005), "Causation, Physics and the Constitution of Reality: Russell's Republic Revisited", Oxford: Oxford University Press.

Savitt S., (2006), "Being and Becoming in Modern Physics", The Stanford Encyclopedia of Philosophy, http://plato.stanford.edu/entries/spacetime-bebecome/

Varela F., (1995), "The Emergent Self", chapter 20, excerpted from Brockman J., "The Third Culture: Beyond the Scientific Revolution", Simon & Schuster

Yates, J. (2008), http://cogprints.org/6232/ , "Experimental philosophy and the MBI", PHILICA.COM, Article number 139.

Yates J., (2008) , "Towards a Science of Consciousness", p147-8, April 8-12, 2008, Tucson Convention Centre, Tucson, Arizona, Center for Consciousness Studies, University of Arizona. This also describes the use of the MBI ("Many Bubble Interpretation") to finally resolve the Schrodinger cat paradox.

Yates J., (2009), "The Many Bubble Interpretation, externalism, the extended mind of David Chalmers and Andy Clark, and the work of Alva Noe in connection with Experimental Philosophy and Dreamwork", http://philpapers/org/archive/YATTMB.1.pdf and on website

Yates, J. (2010), "Not Even Wrong - a view of current science of the mind", http://philpapers.org/archive/YATNEW and on website

Zimmerman D., (2007), "The Privileged Present: Defending an 'A-Theory' of Time", in Hawthorne, Sider and Zimmerman, eds., 'Contemporary Debates in Metaphysics', Malden, MA: Blackwell

Zimmerman D., (2010), "Presentism and the Space-Time Manifold", http://philpapers.org/profile/35 as on 26/11/2010

2.04 *Evolutionary Robotics and its application to time travel*

Our recent paper "Simple Robotic Simulation of Perception, and Time Travel" should eventually appear on our website and it can be found in our blog of March 21st, 2011.

Precise graphical plots of the movement of these machines is still being made, to create a family of curves under various circumstances and for various configurations, as well as for varying datings of internal robot time.

One following approach could entail the deciding of which of these configurations can give the most optimum results. Simple mathematical neural network theory has already failed to give any definite real promise in other connections. So many obvious examples have shown its basic failings, the typical case, perhaps, being in the case of calculations for derivative instruments in the stock market. That is an interesting case as, whilst simple mathematical neural network theory should conceivably be able to take into account most contingent circumstances, just as weather predictions can be expected to lead to some positive results from very simple macro measurements including wind speed, temperature, barometric pressure and so on, perhaps over a very wide (or even global) area without a lot of micro understanding, leaving aside the inevitable problems of chaos theory and the like, some results of a plausible nature may be obtained for weather predictions. But generally we can still think of Lorenz (even allowing for the fact with Cray computers etc. we have gone some way past Lorenz's Royal McBee computer), or on a practical basis be aware that most people have little faith in newspaper weather predictions, for example. But for the stock market, even agreeing that macro parameters should give perhaps at least a rough indication, generally speaking it is no easier than predicting winners of a horse race or results within a cricket game, simply because of fudging by the jockeys or players in ways which the macro parameters might well be, in essence, incapable of prediction.

Even in these simple cases, it can be hard or impossible to predict which banker or jockey or cricket player will do the fudging, or indeed whether it may be some entirely extraneous factor quite outside the scope of the state of play and even outside the metric of the statistical measurements

This does not prevent us from using evolutionary robotics, together with neural networks (Nolfi, 1994).

The simple message at the moment is that we are not currently trying to 'predict the future' results, but rather to establish a working McTaggart A series format for description. Our earlier working experiments used the real results and experimental methods of Gregory (1970), where he used visual techniques to observe astronomic observations - which related to signals from objects now in our own past (Yates, 1984).We evolved the techniques of Gregory so they would work with electroencephalographic measurements in our own laboratories and with the results of Kornhuber et al. etc and our own readingd, and achieved a working time machine.

Therefore at a later point we may use, at the Institute for Fundamental Studies, the Information Field Theory of Torsten Ensslin (Ensslin, 2009) and the help of Feynman diagrams to carry though some of the mathematical integration difficulties.

It is hoped to be able to duplicate the effect of A series time travel, not necessarily with a live human being as we have already done in the way of a psychological experiment, but in a crude way in a very simple system by the use of very basic swarm robotics. Changizi has already pointed out that a simple and somewhat similar B series effect is possible even with existing biomeasurements.

Miglino et al (2008) describe a simple new approach in Evolutionary Robotics according to which human breeders can become involved in the evolutionary process. They use their simple "Breedbots" which is not very different to our own NXT robots (Yates, 2011) or indeed to the usual Braitenberg type 2 robots. That is almost like the creation of an experimental philosophy or psychology interface with non human participants and as such, also has some theoretical interest.

So now we need to redesign a testing plan to see how accurately our robots can obtain type A McTaggart behaviour. This is not the same as building robots which will to all intents and purposes predict the future, although these robots should have the qualities of relevance and experience to Type A behaviour.

The process is likely to be to set up a course which may contain puzzles or obstacles and to see which variations in the robots need to be altered, using evolutionary neural nets, to create robots with the response that we require. We

59

may judge performance by our own subjective judgement as to which robot qualities were successful, or allow a neural network or other mathematical techniques to decide. Tests may need a swarm of robots or the compilation of results by a sequence of modified robots.

Winfield (2011) admits very fairly and reasonably that "Right now we just don't know how to design a system that produces complex overall behaviours from a group of simple agents" so as usual the Institute for Fundamental Studies will start as simply as possible, hoping for results from just one modified model, perhaps with derived videos obtained using the model in several measured configurations - producing a single video with perhaps six approximate clones appearing within it, following different paths which have been measured using an isolated robot to obtain responses as varied by model modifications. The appropriate sets of rules can then be added to each of the six slightly different clones. Then the relative successes of the clones can be assessed from viewing the video which will show how a simulation using all the six in the same video succeeds. To be quite fair, I have looked at many of the swarm robot videos on Youtube, today April 15th, 2011 and they do seem to be unsatisfactory and unfortunately sometimes little more than child's play, with no real advantages. Clearly it is a difficult task to show or prove anything significant at all, but here at the Institute for Fundamental Studies we at least have purpose, method, and current knowledge and we are doing much better then many others in our opinion.

The above work should form the basis of a further note.

References

Ensslin T.A.,Frommert M., Kitaura F.S.,(2009), "Information field theory for cosmological perturbation reconstruction and nonlinear signal analysis",Phys. Rev. D 80, 105005

Gregory R.L. (1970),"The Intelligent Eye", Appendix B, 170 et seq., Weidenfeld & Nicholson, London

Miglino O.,Gigliotta O.,Ponticorvo M, Lund H.H.,(2008), "Human Breeders for Evolving Robots", Artificial Life and Robotics , vol. 13, no. 1, pp. 1-4

Nolfi S.,Floreano D.,Miglino O., Mondada F., (1994),"How to evolve autonomous robots: different approaches in evolutionary robotics", in R.A. Brooks, P. Maes eds., Proceedings of the IV International Workshop on Artificial Life, Cambridge, MA, MIT Press

Winfield A.,(2011), http://www.ias.uwe.ac.uk/~a-winfie/ , http://www.newscientist.com/article/dn13244-shapeshifting-robot-forms-from-magnetic-swarm.html

Yates J.,(1984) Patent Number:GB2051465 Publication date:1981-01-14. I also mention and apply Gott's comment to this patent in http://philpapers.org/archive/YATASO.1.pdf

2.05 Zen and the art of paper aeroplane manufacture - preliminary thoughts only

For awhile during the 1950s to 1980s we were all assailed with articles and books with titles like "Zen and the art of ...". By now there are more than 200 books with titles beginning "Zen and the art of ...".

It seems that the spate was actually begun by "Zen and the Art of Archery" by Eugen Herrigel. That book contains often reasonably accurate ideas about motor learning and control. These may be said to give rules for learning any sport or physical activity. For example, a central idea in the book is that through years of practice, a physical activity becomes effortless both mentally and physically, as if the body executes complex and difficult movements without conscious control from the mind.

The most famous of the "Zen and the art of ..." books was perhaps Robert Pirsig's "Zen and the art of motorbike maintainance" which sold 4 million copies in some 27 languages. This book appears to have been written around 1972 when Pirsig seems to have become acquainted with details [1] of Zen, a subject which was of popular interest to many a good deal earlierin the 1950s and of course had been proposed by Pirsig shortly after his evident catatonia (or possibly Zen enlightenment around 1961) and the subject had been discussed with him during his involvement in the war in Korea in 1947. This book has always seemed to me to have had a great deal to do with Zen, and indeed motorcycle maintainance, although many people, possibly including Pirsig, construe it to be a review of

61

Pirsig's so-called "metaphysics of quality". Pirsig seemingly tries to explain that he seeks a middle ground between rational understanding and romantic perception.

Particularly intriguing is the so-called "Pirsig's Paradox" (Appendix).

In some ways one could glibly argue that all this is something like a union of the McTaggart A series and the McTaggart B series. But as, hopefully, we have seen, it is not quite that simple. For example the work of Alan Watts [8] seems to actually try to involve both rationality and contemplation into the Zen concept.

Then again, Pirsig appears to have been mad and Watts a supporter of the use of hallucinogens, and both approaches are clearly at variance with present efforts.

While a consideration of such views persist, up to a point modern sceptics can readily rebut Pirsig's views [2], perhaps along the lines of reasonable reification arguments. But we bear in mind that our aim is not to establish a view on scientific materialism, or atheism or anything along those lines.There are many practical problems with the (metaphorical) 'new Turks' or Dawkins style views on atheism, as for example in the idea of a higher intelligence such as some form of AI++ [6] or maybe simply the notion a quite likely [7] more advanced entity which must in many ways be equivalent to or even coexist with even a quite banal God concept. That idea is almost like a very basic view of Hinduism! In the present context one tries to stand back from such contests and there seems to be an ultimate irony in the idea of atheists attacking a godless religion like Buddhism. But such argumentation is, at most, a side issue. Pirsig seems to think he is saying something and we are trying to extract a useful root view from it, or relate to a useful root view. We basically could appear somewhat like the 'six blind men of Indostan' in the John Godfrey Saxe poem about the six blind men and the elephant. Rovelli [3] may say "everybody hears everybody else stating that they see the same elephant he sees" but it is far from clear that even that is the case and we can easily find that we are in some kind of philosophical regress. This is not our desired territory and we do not want to juggle with words either.

So we presently neither want to seek some rebuttal, nor are we intending to promote a view on medical or non medical drug applications. Rather we are studying time and consciousness, often in terms of the A series and the B series.

We therefore consider the relation of modern neurophysiology to Buddhism [9, etc] and also the waterfall effect, the McCollough effect, the reverse-McCollough effect, and so on, and get as close as down and dirty as we can with paper aeroplanes [4] and similar modern ideas which may be of a similar nature in application to Zen archery and contemplation. This is less to improve mental or physical agility but more to improve and re form our ideas and concepts.

We note also the well-intentioned thoughts of Adrian Bejan and his constructal ideas, "Cognition is the name of the constructal evolution of the brain's architecture, every minute and every moment," Bejan said. "This is the phenomenon of thinking, knowing, and then thinking again more efficiently. Getting smarter is the constructal law in action."

Here, too, the work of Prigogine and his believed temporal entropic direction ideas may be considered, as thermodynamics is also being implied by Roy Baumeister as being directly related to free will and decision making. Here our own Romeo and Juliet concepts [5] using Sprott's ideas may help even to introduce an implied mathematical formalism.

Appendix

In "Zen and the Art of Motorcycle Maintenance" page 38, Pirsig writes "(ghosts) are unscientific. They contain no matter and have no energy and therefore according to the laws of science, do not exist except in people's minds. Of course, the laws of science contain no matter and have no energy either and therefore do not exist except in people's minds. It's best to refuse to believe in either ghosts or the laws of science." Pirsig then explains that he does believe in ghosts. "you will go round and round until you realize that the law of gravity did not exist before Isaac Newton. So the law of gravity exists nowhere except in people's heads. It is a ghost!"

References

(1) http://www.psybertron.org/timeline.html

(2) http://www.skeptic.com/eskeptic/10-04-28/

(3) Smerlak M., Rovelli C., http://arxiv.org/abs/quant-ph/0604064

(4) http://www.sciencetoymaker.org/hangGlider/index.htm

(5) http://ttjohn.blogspot.com/2009/08/explorations-of-available-philosophical.html#links

(6) Chalmers D., http://www.consc.net/papers/singularity.pdf

(7) http://cosmiclog.msnbc.msn.com/_news/2010/06/15/4512943-an-avalanche-of-alien-planets

(8) Watts A, (1973), http://video.google.com/videoplay?docid=-17193013653573213313

(9) Eskandari P, Erfanian A., (2008), "Improving the performance of brain-computer interface through meditation practicing", Conf Proc IEEE Eng Med Biol Soc. 2008;2008:662-5 ,
http://www.ncbi.nlm.nih.gov/pubmed/19162742

2.06 The future beyond Sections 2.02 and 2.04

There are many problems involved in going beyond the very simple approach being followed in Sections 2.02 and 2.04 .

Minsky's "third interesting possibility: psychology could turn out to be like engineering" really is not enough, nor is Chomsky's "physics-worship". This is clearly because both workers are both basically just running down an escape tunnel, that of the B series. We must at least consider the A series.

So, if you like, we cannot find totally acceptable the so-called 'computational theory of mind'.

The mind is not really a 'society of agents' nor a 'a large collection of partly finished drafts'

Or to put simply, the Blue Brain project of Henry Markham and others will not work as expected. "Steampunk" is the nicest word I can use to describe Blue

Brain results to date, but "retro" would be more accurate with my current understanding, as "digital" and the like actually do seem to lack many qualities of so-called "analog". Whilst "wet lab" techniques may certainly going to be necessary in such work, and it is a step forward that they are using them, even a simple biocomputer is unlikely to produce early required results.

Balakrishnan's (2001) book of papers should provide a helpful signpost but Dautenhahn pointed out "similarly to a child who might pretend that the wooden stick in his hand is a sword but clearly knows that the object is not a real sword,we *know* that computers are not people".

Now Dautenhahn clearly tries to get to the nub of the matter in his various writings, in Dautenhahn (2009) and particularly for example in Dautenhahn (2007).

At the moment, to relate all this to VR studies of time is probably a relevant aspect of the present approach. We have always born in mind the possibility of further use of virtual reality for cyberstimulations and direct brain stimulations, And since the present study primarily involves time, the possibility of robot life and of robot intelligence of human comparability need not enter into the matter unless it has to.

I think now we at least know what we probably CANNOT do and that is to create, at least in the simple ways so far suggested by Minsky and others, some kind of golem.

And that likelihood should in itself be of value in further studies. And so far we have sketched out a plan of what at least may be possible with the resources at our disposal. What follows almost immediately is that thinking biocomputers, unlike electrical or mechanical ones, sound at least possible as a fact. One can almost readily imagine a mouse made synthetically, for example using genetic components and sections from a 'real' mouse, but the equivalent mechanical appliance would be hard to believe as capable of real thought. But in our present studies it really does not matter and that fact - the fact that we are not actually trying to create consciousness but merely are doing time travel experiments, though keeping a sharp look out for any other worthy results - currently seems to be of merit. So we can certainly use biocomputers in our present studies and it does seem more likely that they will lead also lead to a development of consciousness much more easily than the present steampunk electronic computers. Part of the rational argument for that is importantly that they should involve the A series as well as the B series. We are trying hard to get away from the deeply embedded philosophy of Leibniz or of Newton, and it is not easy.

References

Balakrishnan K., Honavar V., Patel P M.J., (2001). "Advances in the Evolutionary Synthesis of Intelligent Agents", Several articles especially including Michel O., p185 MIT Press, 2001.

Dautenhahn K., (2007), *Phil. Trans. R. Soc. B* 2007 **362**, 679-704

Dautenhahn K., Nehaniv C.L., (2009), "Imitation and Social Learning in Robots, Humans and Animals: Behavioural, Social and Communicative Dimensions. 1st ed., Cambridge University Press

3

Dreams & the Reverse Stickgold Effect with the MBI

Section 3.01 The idea of dreams being mere internal artifacts of the mind does not seem to be essential to externalism and extended mind theories, which seem as if they would function as well without this additional assumption, once the MBI can be invoked.

Section 3.02 discusses backward leakage in dreams and introduces the Reverse Stickgold effect experimentally.

Section 3.03 discusses and up to a point agrees importantly - having some reservations however - with Hobson's view that dreams are "the royal road to consciousness".

Section 3.04 points out that experiments along the lines we are doing, can bring results and indicates some further lines of research.

3.01 The Many Bubble Interpretation, externalism, the extended mind of David Chalmers and Andy Clark, and the work of Alva Noe in connection with Experimental Philosophy and Dreamwork

Abstract

The idea of dreams being mere internal artifacts of the mind does not seem to be essential to externalism and extended mind theories, which seem as if they would function as well without this additional assumption. The Many Bubble Interpretation could allow a simpler rationale to externalist theories, which may be even simpler if the assumption that dreams have no worthwhile content outside the mind is omitted.

Almost everybody agrees that mind and world are causally coupled (Prinz, 2009).

In normal perception we don't have the problem of stabilising detail. Noe (2009) points out - and in that follows the point made by LaBerge and many others - that if a dreamer looks more than once at, say, a printed sign in a dream the sign is likely to say something different on second viewing (apparently always in LaBerge's experience). The sign's content may even change whilst it is being watched. It often does in my experience of dreamwork. In general terms I have found that recorded results of dream experiments are more consistent over a series of experiments involving many different dreamers, than some experiments in the waking world which could be expected to be much more easily quantifiable and even more easily measurable (Yates, 2008). Synaesthesia is one example. That is to say with dreamwork we are left with a pile of reasonably consistent data to consider, even though the data itself may be construed as irrationally produced or even arising from a random source. Others such as Domhoff seem to confirm this.

But, even though dream results are in some way scientifically collatable, there is still the problem with instability of detail within individual dreams. For example, it is not like measuring UV spectra, when the same pure substance should give the same results each time. Now variability is not too unusual in psychology experiments, but clearly this level of variability is well beyond the norm.

Noe interprets this as meaning that dreams are not real, in the sense that waking experience is real, though he admits that the perceptual experience of dreaming is real. He then reasonably says that this implies that waking experience is different to dreaming experience, and that dreams cannot be construed as evidence in effect that reality is just another dream.

Yes, we can go this far. Andy Clark and Chalmers also seem largely to admit this view. In principle for day to day working purposes we can accept such an idea but we would use Ockham's razor yet again to say that it does not mean, either, that dreams necessarily consist ONLY of results from within the brain. In a way Noe would betray the thread of his own argument if he took the view that dreams necessarily consist only of results from within the brain, as he is very much into Merleau-Ponty type externalism.

In other words, if we accept Noe's variety of externalism in principle, we should leave it as open ground that dreams may come not just from a simple B-series 3+1 dimensional lump of white and grey neural matter. In fact our own position involves the A-series as well, where externalism, we trust, provides less of a problem. (Yates, 2008, 2008a, 2008b).

So such a position should be arguable and indeed essential at least in principle with Noe.

And in fact the mental status of dreams is still not clearly known, and there is no need for dreams to be purely internal in origin to allow most of Clark's argumentation.

69

Of course anyone who seems to claim entirely the unfortunate implied position that we only need consider what goes on in the brain to understand how the brain operates in the world, probably can be dealt with using a slightly different paradigm, not the topic of the present note.

The position of both Andy Clark and of David Chalmers seems to differ quite substantially from that of Noe. Although they both seem to favor augmented extensions of the simple B-series 3+1 dimensional lump of white and grey neural matter, it is sometimes difficult to visualise a satisfactory precise detailed formulation of that idea. Without a clear A series somewhere, instinct tends to make one fall back to Fodor's position.

With the A series on board, things differ dramatically. Purely to illustrate this point in another quite different case, if we consider Butterfield's (2001) critique of Barbour's work (Part 3.1), we realise that the moment McTaggart's paradox is invoked, the situation changes drastically.

It looks to me that Andy Clark is apparently activating the Foundation Argument against dreams, according to his correspondence with Noe (2009) at least. This seems to be a needlessly blunt edged sword to establish his views. The idea that consciousness depends only (or mainly) on what is happening in the brain the brain would take him away from the more extreme stance of Noe. And the Foundation Argument idea seems quite unproven anyway. It is essentially close to the idea of Crick. Noe gives quite cogent reasons against that, which may not concern the present argument. It really hardly matters to Clark's main argument that he should also apply it against dreams if he applies it to a lot of other things too. In other words, Clark's occasional comments that "dreams" seem to constitute almost purely cranial/neural matters do not really seem to be a clincher to his argument, but incidental ideas.

And as I just pointed out, we can look at even more unfortunate paradigms elsewhere. To put it largely, arguments against solipsism and generally accepted brute fact can come into such a discussion, except in specialised areas of mind science. There we are talking of Churchland as well as Crick and Searle, and our own working argumentation would be presented differently in such a case.

So it is perhaps simplest to look at the views of Fodor (2009) and Chalmers (2009). Fodor is claiming that Clark is effectively using the slippery slope argument between Otto and Inga, and also that notebooks are not the same as minds. Now that, apart from intentionality factors, does seem to be the case.

Time Travel in Theory and Practice

I'm not sure of the dispositional claims of Chalmers. Obviously Chalmers seems maybe to eventually want to solve a problem which Noe refers to on the first page of his preface "Only one proposition about how the brain makes us conscious . . . has emerged unchallenged: we don't have a clue." The Chalmers solution could presumably involve yet more machines, and Fodor presumably would not quite take that line.

I think here we are left with different levels of implied logic. Noe's approach could be construed as a sophisticated "ad hominem" level argument, briefly to the effect that everyone knows that the world exists in the way most people think it does (for example, in no sense is it anything resembling the Matrix). That is somewhat in the sort of way that some have said Dr. Johnson tried to refute Bishop Berkeley, by kicking a stone and saying "I refute it thus". The problem is of course not just time constraints but more importantly constraints as to apparently available accurate scientific knowledge about situation and circumstances. In the context of neurophysiology and consciousness research generally we frequently seem to be deep into the area of informal logic (Groarke, 2008, 1999). Of course Noe has written a lot to expand and rephrase his arguments and to include a great detail of neurophysiological detail, and that is indeed of great value to his comments, and that fact must be remembered.

But in the case of Clark and Chalmers I am left feeling that they are looking for more of a logical commitment (Groarke (1999), Walton, (2002)). The most recent post of Chalmers (2009) suggests that it has not got there yet, but my thinking is that the sort of overall approach that Mandik (2009) has used, where he actually goes so far as to question the current idea of representation, is perhaps more relevant. Details in Mandik's case are sketched out by others in Mandik (2009a).

Whether there is an appropriate formalism is not the point here, as dreams is our current topic in this note. But it is possible to point out that it is only if the brain is considered as a simple B-series 3+1 dimensional lump of white and grey neural matter that worries about externalism overawe us so much. In the Many Bubble Interpretation (Yates, 2008, 2008a, 2008b) the relationship comes out naturally. Simple mathematics is not there in detail yet but so far all seems straightforward.

To look briefly forward, purely for a simple modus operandi in experimental philosophy, I consider that Knobe's (2009) style of approach may be better than a lot of mathematics before we can contrive more parameter values.

Conclusion

It cannot be assumed that dreams are of necessity simply part of a simple internal mental continuum.

The Many Bubble Interpretation could allow a simpler rationale to externalist theories, which may be even simpler if the assumption that dreams have no worthwhile content outside the mind is omitted.

References

Butterfield J., (2001), "The end of Time ?", arXiv:gr-qc/0103055 v1 ; especially for example 3.1

Chalmers D., (2009), http://fragments.consc.net/djc/2009/02/fodor-on-the-extended-mind.html

Fodor J., (2009), http://www.lrb.co.uk/v31/n03/fodo01_.html

Groarke L., (2008), "Informal Logic", The Stanford Encyclopedia of Philosophy (Fall 2008 Edition), Edward N. Zalta (ed.).

Groarke L., (1999), "The Fox and the Hedgehog: On Logic, Argument, and Argumentation Theory", ProtoSociology, (13), p29.

Knobe J., Phillips J., (2009), Psychological Inquiry, Volume 20, Issue 1, 30 - 36,

Mandik P., (2009a), http://www.consciousentities.com/?p=117

Mandik P., (2009), Journal of Consciousness Studies, 16, No. 1

Noe A., (2009), "Out of our Heads", numerous pages: including p179 on dreams, p203 on Andy Clark etc, ; Hill & Wang.

Prinz J., (2009), "Is Consciousness Embodied?" [In P. Robbins and. M. Aydede (Eds.) Cambridge Handbook of Situated Cognition. Cambridge: Cambridge University Press (forthcoming)] ; http://subcortex.com/IsConsciousnessEmbodiedPrinz.pdf

Walton D.N., (2002), "Are Some Modus Ponens Arguments Deductively Invalid?", Informal Logic, Vol. 22, No. 1 (2002): pp. 19-46 ; http://io.uwinnipeg.ca/~walton/papers in pdf/02modus.pdf

Yates, J. (2008a). http://cogprints.org/6176/ , "Category theory applied to a radically new but logically essential description of time and space",PHILICA.COM, Article number 135.

Yates, J. (2008b), http://cogprints.org/6232/ , "Experimental philosophy and the MBI", PHILICA.COM, Article number 139.

Yates, J. (2008). "A study of attempts at precognition, particularly in dreams, using some of the methods of experimental philosophy." , Philica.com , Article number 146.

3.02 Do we Dream of the Future ?

Abstract

Stickgold's Tetris experiments were repeated with variations to allow the possibility of category theoretic backwards leakage to be observed. Backwards leakage in dreams, sometimes referred to imprecisely and in a rather more semantically loaded way as precognitive dreaming, seems to have occurred.

Introduction

Blackmore's (2002) paper is possibly the most up to date detailed account of controlled trials on precognition. Her remarks go back as far as the early dream work of Kilbracken. Any results reported in Blackmore (2002) or implied by it seem to suggest that the subject does not present much future hope for precognition. Specifically with regard to dreams, Hobson (2006, 2005) doubts if there is any precognitive element in dreams though he seems to have had at least one dream which could be fitted to that category (a totally different thing, of course). I am largely in agreement with Hobson's position on interpretations to date though there is still much exciting work to do, some of which I begin in this essay .

There are of course a great number of papers in the popular literature. For example Obringer (2006) gives a detailed 'popular' account with many citations and links, Stowell (1995, 1997, 1997a) has some older literature and Ben Goertzel mentions some ideas in his journal "Dynamical Psychology" which appears mainly on the web. By and large Hobson's and Blackmore's contributions, at any rate, should not be disregarded.

So generally speaking Hobson's and Blackmore's results and the largest fraction of similar work carried out with reasonable scepticism suggests that precognition does not happen, inside or outside of dreaming, under the constraints and conditions which have been imposed to date..

However we have already pointed out in earlier essays in this blog (see references at end) that existing statistical methods may be inappropriate here and also that there is much more to be found or interpreted through observations.

The Experiment as Performed

This experiment helps us to discover the degree if any of backward leakage using the present category theory approach which can be obtained by dream studies. The category theory approach does not stand or fall on the basis of such studies but they could certainly provide a pleasant confirmation of its correctness. It may ultimately replace the present crude 19th/20th Century idea of punctal time, which seems like only a dim shadow of reality in comparison.

Stickgold (2000, 2003, 2005) controlled the content of 17 different people's dreams after the first hour of sleep. Twenty-seven test subjects played Tetris on Nintendo sets for three days, with a two-hour morning session and a one-hour evening session the first day, and a one-hour morning and evening session the following days. Of the 27 people, 12 were beginners to the game and 10 were experts. Five of them were amnesiacs as well.. Seventeen members of the group recalled dreaming of falling Tetris pieces at least one hour after falling asleep. Most of the dreams occurred the second night.

As far as I am aware these are the first major trials in history which so consistently seem to induce specific and definite dreams, and as such they should be extremely relevant to the present backward leakage study. The problem with psychological tests as compared to simple physics measurements has usually been the great difficulty in obtaining clear and consistent results. The chequered history of, for example, the Perky effect which I mentioned in detail earlier is an extreme but only too typical example. At the other end of the scale we have, say, Milgram's (1974) torture experiments which seem to have had very high repeatability all over the world.

Therefore we first did simple tests with just one subject (subject X). I do not recall dreams frequently so I was simply an observer. An abbreviated process was used during which from Night 1 to Night 9 the dreams were simply recorded with particular reference to anything like falling Tetris pieces, from Night 10 to Night 23 subject X played space invaders or other games without Tetris shaped pieces, from Night 24 to Night 26 the subject was also primed with 3 Tetris demo games (though was instructed to casually observe these but not able to play, and subsequent to this was asked to play the usual number of games of space invaders). From Night 27 to Night 29 the subject played Tetris and recorded the score and from Night 30 to Night 40 no games were played.

The results are given below in tabular form. SI means "Space Invaders", T means "Tetris", TD means "Tetris demos but no play" and N means "No game". DSI means "Dream reflecting Space Invaders or alternative games to Tetris" DT means "Dream reflecting Tetris", DO means "other dreams recorded", DN means "no dream available (or dreamed)". A query appended means ambiguity, usually only slight ambiguity.. We have more details than the brief ones below, and of the other games unlike Tetris played but those given illustrate the point made. We have not previously used the subject (subject X) for dreams tests before nor did subject X play Tetris for at least a year prior to the tests. X claims to have been a medium to strong Tetris and Space Invaders player some years ago, and really never to think at all about the games nowadays and this seems a realistic assessment. Attempts were made to have the subject play at least 2 to 3 hours per day of the games as in the Stickgold trials.

Night	Game Played	Result
1	N	DO
2	N	DN
3	N	DO
4	N	DN
5	N	DO

6	N	DO (often fragmented), DT (white pieces)
7	N	DO (often fragmented), DT (boxes appeared like lights)
8	N	DO
9	N	DO
10	SI	DO, DSI
11	SI	DO
12	SI	DO, DSI
13	SI	DO, DSI
14	SI	DO, DSI, DT?
15	SI	DO
16	SI	DSI, DT?
17	SI	DSI, DSI
18	SI	DO, DSI, DT?
19	SI	DSI, DO
20 (SIMULTANEOUS)	SI	DO, DT/DSI
21	SI	DT/DSI? (uroboric symbolism?)
22	SI	DO, DT/DSI, DT
23	SI	DT/DSI (MORE T THAN SI)
24	SI, TD	DO, DT?
25	SI, TD	DO/DT?
26	SI, TD	DO, DT?
27	T	DO/DT?
28	T	DO, DT
29	T	DT/DO
30	N	DT/DO
31	N	DO
32	N	DO
33	N	DO
34	N	DO/DT
35	N	DO/DT
36	N	DN
37	N	DO
38	N	DO
39	N	DO

There were thus up to at least 12 Tetris dreams on 12 nights prior to the actual playing of Tetris, by a person with no interest or thought of Tetris or of video games. The above results therefore show a clear backward leakage effect (or what a tyro might interpret as precognition) for both Tetris and, to a lesser extent, for the Space Invaders style games. Probably many more experiments, with additional subjects, need to be carried out to establish this effect with certainty. It should be warned that no clear interpretation, whether of a positive or a negative nature, is going to be easy, but early results look promising. Yet again, I remind of the tricky nature of this work. As an example, I see Hobson's 'rejection' of precognition, after giving an example of a 'precognitive' style dream - and on grounds of simple statistical merit I wholeheartedly confirm agreement with his view. The fact remains: Hobson had at least one 'precognitive' dream and by altering experimental circumstances I have very easily encountered probably more than another 10 'precognitive' dreams. And I do not believe in 'precognition' either, but there certainly are some interesting effects out there to study. I will not wager any bets as to my interpretation of further 'independent' trials.

Appendix A "Two time dimensions ?"

Do we really need to use category theory to achieve our results? My answer would be "Probably, if we want to get a result we can in some way easily understand, anyway".

An alternative approach might be, effectively, to have two time dimensions in (normal) punctal time.

Feynman, almost needless to say, apparently tried to devise such a world with two time dimensions in henry.pha.jhu.edu/Henry.Feynman.pdf but did not follow it up further though he found the idea interesting.

Wikipedia and other such sources also have their own take on more than one time dimension, as for example in http://en.wikipedia.org/wiki/Second_Temporal_Dimension.

Time Travel in Theory and Practice

Astronomy-cafe give a much more basic approach.

At http://www.astronomycafe.net/qadir/q1980.html they say: "What would a universe be like with two time dimensions? Pretty weird. I do not know the details, but because there are two time dimensions, it is very difficult to avoid creating 'loops' in which one traveller leaves location A and travels to location B and arrives at a specific time in the future, and another traveller making the same journey but arriving at 'B' before he left 'A'. Presumably, we would have the same freedom to move in this second time dimension as we do in a second space dimension, and because closed curves can be created in 2-dimensional spaces, the two time dimensions would allow us to 'move' along paths that violate causality. Every pair of points in space would be threaded by a multitude of such possible curves, and the concept of cause-and-effect would become meaningless"

Foster (2003) did a PhD thesis on a system with two time dimensions which is available for reading and to which I give a reference. This could be an alternative approach, still using punctal time. But I would imagine that to use punctal time would be the equivalent of using epicycles to explain an earth-centred solar system, that is to say it might do for some aspects to provide a link, an interpretation, an understanding - I say this, bearing in mind only such factors as the long run that the Copenhagen Interpretation of quantum mechanics had, or indeed the idea of the earth being at the centre of the Universe. But the diversity from fact here, seems much wider and is of a theoretical nature, not simply an explicatory one. For the moment, we will stick with what we have.

Gott (2001) used the term "dream time" or "dreamtime" (a term used by native Australians in what is normally taken to be a mythological sense) to describe a second time dimension and pointed out the possible analogy or comparison with time states during dreams, but, like Feynman, did not elaborate the model though he pointed out the fact that the unbridled use of a second time dimension could make comparison of such a model with the real world unrealistic.

Appendix B Can simple logic avoid the use of category theory?

Can simple logic (Lob logic or indeed any kind at all) avoid the use of category theory? My answer would be: "Simple logic without bowing the head to McTaggart only makes things worse, not better. And if we admit McTaggart we probably need category theory. We probably really do need category theory, anyway ".

I consider for example the Ostrogorski paradox and what it entails. Pigozzi (2005, 2006) expounds on the Ostrogorski paradox in some detail usefully and in logical terms.

A brief non technical example from Widerquist (2003) may simply illustrate the matter, as follows "The Ostrogorski paradox shows how the fulfilment of the majority's preferences can make everyone worse off, such as electing a candidate who holds a minority opinion on all issues [Kelly, 1989; Nermuth, 1992, *op. cit.*]. In one version, three voters use the majority rule to redistribute income. Each of the three possible majority coalitions votes to take two dollars from the third voter and redistributed it among themselves, but one dollar is lost in transactions costs. Thus, although each decision benefits the majority that votes for it, every voter is worse off than they would have been if none of the votes had taken place. This example involves pure rent seeking, and cannot happen if people are unselfish. Something very similar, however, can happen. Milton Friedman [1962] used an informal example in his book 'Capitalism and Freedom' that was in fact an Ostrogorski paradox motivated solely by concern for the public interest. He argued that many different majority coalitions could be persuaded to ban many other kinds of speech and nearly everyone would find herself wanting to say something that had been banned by the majority. Thus, it would be possible, he hoped, for a majority coalition of voters to see that it is in the interest of society to ban the banning of any kind of speech.

The Ostrogorski paradox may be more difficult to resolve when individuals are altruistic than when they are selfish. Tullock [1984] argues that intellectuals could make a real contribution by pointing out the costs of special interest legislation, and hopefully the bulk of such legislation would not pass. Economists have made efforts for many years to point out these costs with little success. If voters are motivated purely by self-interest and these government actions are inefficient forms of rent seeking, it would be in the interest of everyone to form a broad coalition to ban all such government actions. Why has no such coalition formed?" And of course I am talking about statistical analysis not political coalitions, but real political coalitions involve specific votes or voting patterns and statistical analysis involves comparison of measurements so there is quite a direct parallel. I am not simply looking for an analogy, but for physics. But in the early stages we have to walk warily. A simple statistical analysis of dreams and their contents along the lines of Domhoff's work is not going to have a one to one mapping onto future or present scientific results. Certainly Domhoff does not seem to claim in the circumstances he is dealing with that it will, and we will not come up with dream prediction statistics that way either.

Another case: Bacchus (1989) and Chau (2006) also illustrate non-transitive effects which are found everywhere, for example in the children's game "Scissors, Paper or Stone ?" and indeed in betting wagers such as those surrounding the Penney-Ante 'paradox' (Andrews 2006). There could be plenty of food for thought for further work in those examples.

The factors above do apply to statistical methods for correlating and computing psychological effects during dreams (as perhaps unlike those of the Stickgold priming experiments done slightly after dreams and prior to dreams for priming. During dreams these effects may well apply in a different way). Here we are trying to statistically 'average' results not readily susceptible to averaging. For example we cannot simply 'combine' or 'count' results on different peoples dreams. It is not just a matter of statistical correlation coefficients or something of the type but of trying to avoid giving entirely different meanings to phenomena. e.g. One persons dream of a certain horse winning a race may have, in the scope of different person's day to day activities, have a totally different meaning to another's. e.g. a sporting or racing enthusiast may have one interpretation and an artist with personal problems have another, and a small boy still another. That is not to say an overall slant or impression can never be made but it certainly may not be easy or mathematically mindless, as statistical correlations frequently try to be for simplicity and/or objectivity.

So the idea of "dreaming racehorse winners or share prices" is not a testing exercise, it is simply generally outside our remit as a day to day possibility.

Also we must remember here that, for a simple statistical analysis we are (or at least expect to be) essentially also very far from punctal time or the time of Einstein/special relativity. We briefly remarked on that in the two earlier Varela essays, "Neurophenomenology and Category Theory" I and II. In Gamow's "Mr. Tompkins in Wonderland" we see how very easy changes in physical parameters can change the conception of parts of the universe markedly and of course in this present case we only have category theory and McTaggart to fall back on, not a very easy 4D space or even some string theory manifestation of much the same thing.

But it is in the light of the above paragraphs and the earlier work in the blog that we can legitimately try to interpret whatever relevant results we obtain.

We are not totally in the field of the unknown and we can interpret whatever reasonable results we can obtain. But it may not be easy. For example a real comparison between our present situation and conventional statistics would be as between a politician's interpretation of a vote on an EU committee (e.g. as in Widerquist 2003) as compared to a simple minded mathematical interpretation of the same vote, using common statistics as would be applied to say a set of elementary readings in an undergraduate physics or even a psychology class. You are looking at very different viewpoints I am not afraid to say. But I have to say that what we are trying seems difficult but not impossible nor outside physics or its criteria. Maybe Popper's approach or Kuhn's approach or some other such approach has as much merit as we can expect on rough new ground. But we cannot, for example, necessarily predict horse races by comparing the results that various people get in their dreams and/or taking some simple average of the results of dreams, nor do we claim to.

Appendix C Philosophical Problems and Specific Details

Many of the philosophical details of this approach have already been discussed in previous blog entries. What will be dealt with here is mainly matters relating to further philosophical matters specifically conserning the dream state and the experiments in hand.

We consider the difference between explicit knowledge and implicit knowledge. Bertil Rolf comments as follow: "The most important type of implicit knowledge consists of representations that merely reflect the property of objects or events without predicating them to any particular entity or event. The clearest case of explicit knowledge of a fact are reflective representations of one's own attitude of knowing that fact. These distinctions are discussed in their relationship to similar distinctions like procedural-declarative, conscious-unconscious, verbalizable-nonverbalizable, direct-indirect tests, and automatic-voluntary control". Many people choose to write in depth about such matters, and the topic is dealt with from various views in the Stanford Encyclopedia and elsewhere.

To explain what I am getting at here it is possibly easiest to quote Bo Newman (2000) who gives a simple example "One thing that characterizes implicit knowledge is that meaning must be inferred. This is because, unlike with explicit knowledge, the codification process is incomplete. Explicit knowledge can be interpreted totally based on content, whereas interpreters of implicit knowledge must rely on some form of previously retained knowledge. The potential for ambiguity is one of the characteristics of implied knowledge. Most readers of the sentence, "Ann put on her heavy coat and locked up her classroom," implicitly understand that it is winter and Ann is a teacher, but there are other inferences that could be made as well. For consistent interpretation, both the person making the statement and the person interpreting it must share some common frame of reference to understand when heavy coats are worn and who locks up classrooms.

Implicit knowledge artifacts can also be found in process-specific software. In developing the software, the designers had to conceptualize the processes that the software would be supporting. That knowledge will then show up in the way the software is intended to be used and in the range of behaviours it directly supports. Even if not explicitly apparent, these implicit knowledge artifacts will effectively constrain users' actions". The profound difference between implicit and explicit knowledge as defined therein makes it clear just how difficult a statistical assessment of results obtained from dreams can be. To go back to the racing dream prediction problem referred to in Appendix B, it becomes particularly obvious that whatever the limits and areas which may well be defined for both explicit and implicit knowledge, these will likely appear to be different for a racing tout and an average small child, for example, and to differ from one individual to the next.

Julian Holley (2004) suggests mathematical models to describe dreamlike cognitive processing. This is a good idea but, especially bearing in mind the above, I am still at a loss to see how an accurate model is to be developed. Briefly "A simple extension is made to the ACS that uses the incomplete information contained in the classifier list as a basis for an abstract world model in which to interact or 'dream'. The abstract thread or dream direction is an emergent property of the selection process, this can be used to recycle around well known states and reduce real world interaction. The system is applied to two simple problems, the random walk and T.maze experiment and demonstrate that they require considerably less interactions with the real world to develop confident world models. Further models and extensions are proposed to advance the system, such as environmental directed generalisation and speculative rule creation." Further progress will be interesting and some of the techniques in that paper could possibly be incorporated in the present model, though if possible it would be best to get some concrete results in the present direction.

Appendix D Further Problems with Results and Observations

The chief problem likely to arise is exactly the same one that seems to have worried Stanley Milgram (1974) in his "Obedience to Authority" program. This is simply that the strength of the authority can determine the results themselves. Milgram tried to get round this by using several different locations for his experiments, for example an office in a run down suburb as compared to a well equipped Ivy League University laboratory and he found that in his case, the very noticeable authority effect which he was measuring was thereby strongly diluted.. Our problem here is greater in that also such effects can work in the opposite direction and the our 'leakage effect' may clearly be not as strong as Milgram's obedience effect. I must say that on beginning these experiments I had some strong doubts about the previous Stickgold (2005,2004,2000) experiments and wondered if even Stickgold's experiments would work at all outside of an Ivy League ambience and was indeed gratified to discover that they did work and not only did we obtain Stickgold's result but we were fortunate enough to apparently go one better. But does it pay to do better than an Ivy League University ? Not really a good scientific question, perhaps, and only time will tell. But there is one serious rider to the situation, and that is that Stickgold's experiments did not seem to cover the case where Tetris was simply observed but not actually played. I thought deeply about priming before I set the experiment up. I was unsure that I would even get the Stickgold result (and am still mildly surprised that I did), and in fact prior to carrying out the experiment I mentioned to Professor Hobson (2006) a slightly more elaborate and much larger scheme which I had had in mind. I still have my doubts as I think that something even more carefully designed (but not necessarily enormously large) is needed. But in fact my point has been made and the ideas probably only require further careful testing, whilst taking considerable care of ambience.

References

Earlier entries in the present blog will give best details for any other queries which may arise about theory and methodology. These entries are at http://ttjohn.blogspot.com/ .

Andrews, M.W. (2006) http://www.gatsby.ucl.ac.uk/~mark/notes/penneyante/

Bacchus F., (1989) A modest but semantically well founded inheritance reasoner. In Proceedings of IJCAI-89, 11th International Joint Conference on Artificial Intelligence, pages 1104--1105, Detroit, MI

Blackmore, S. , Rose, N. , (2002) Journal of the Society for Psychical Research, 66, 29-40

Chau, C. (2006), "Nontransitive Dice Paradox in Networking", http://www.cl.cam.ac.uk/.ckc25/dice

Foster, J. (2003) http://www.phy.duke.edu/ugrad/thesis/foster/

Gott, J.R. (2001), "Time Travel in Einstein's Universe", 62, 66-7, Phoenix, London.

Hobson, J.A. (2006) private communication with author

Hobson, J.A. (2005), "13 Dreams Freud Never Had", 83, Pi Press

Holley, J. "First Investigations of Dream.like Cognitive Processing using the Anticipatory Classifier System", Learning Classifier Systems Group Technical Report UWELCSG04.002
http://www.cems.uwe.ac.uk/lcsg/reports/uwelcsg04-002.pdf

Lange, R. , Schredl, M., Houran, J., (2005), Dynamical Psychology, http://www.goertzel.org/dynapsyc/dynacon.html, "What Precognitive Dreams are Made of: The Nonlinear Dynamics of Tolerance of Ambiguity, Dream Recall, and Paranormal Belief"

Milgram, S. (1974) "Obedience to Authority", Harper & Row, USA

Newman, B. (2000) http://www.brint.com/wwwboard/messages/6818.html

Obringer, L.A., (2006) 'How Dreams Work', http://science.howstuffworks.com/dream.htm/printable (A popular style of presentation. Some of what is talked of in this work refers to a $1.6 million share profit, seemingly not repeated however)

Pigozzi, G. (2006) Forthcoming in Episteme: A Journal of Social Epistemology
http://www.dcs.kcl.ac.uk/staff/pigozzi/pdf/Episteme_Pigozzi.pdf

Pigozzi, G. (2005) "Should we send him to prison? Paradoxes of aggregation and belief merging". In We Will Show Them: Essays in Honour of Dov Gabbay, Vol 2. S. Artemov, H. Barringer, A. S. d'Avila Garcez, L. C. Lamb, and J. Woods (eds.), College Publications, pp. 529-542

Rolf, B. (2004) http://www.nt.fh-koeln.de/philosophyandinformatics/AutumnMeeting2004/sois/SOI_Rolf.pdf

Stowell, M. S. (1995). Researching precognitive dreams: A review of past methods, emerging scientific paradigms, and future approaches. Journal of the American Society for Psychical Research, 89(2), 117-151.

Stowell, M. S. (1997). Precognitive dreams: A phenomenological study. Part II. Discussion. Journal of the American Society for Psychical Research, 91(4), 255-304.

Stowell, M. S. (1997). Precognitive dreams: A phenomenological study. Part I. methodology and sample cases. Journal of the American Society for Psychical Research, 91(3), 163-220.

Stowell, H., Bullock T.H., Basar E., How brains may work : Panel discussion, in Basar E., Bullock T.H. eds. Brain Dynamics, Springer-Verlag 1989, 482-511.

Stickgold, R.J., (2005), "Sleep-dependent memory consolidation". Nature, Vol 437, 27 October 2005, 1272

Stickgold, R.J., Fosse,M.J., Fosse,R., Hobson,J.A.(2003), "Dreaming and episodic memory: a functional dissociation?", J.Cogn.Neurosci.15,1 –9

Stickgold, R.J., Malia,A., Maguire,D., Roddenberry,D.& O 'Connor,M.. (2000), "Replaying the game: Hypnagogic images in normals and amnesiacs",.Science 290,350 –353.

Widerquist, K. (2003) "Public choice and altruism" Eastern Economic Journal, Summer 2003

3.03 Dreams - "A Royal Road to Consciousness"?

J. Allan Hobson suggests that "dream consciousness is ontogenetically prior to waking consciousness and that it serves a foundational function in preparing the brain-mind for its highest evolutionary achievement, waking consciousness in human animals. REM sleep may constitute a protoconscious state, providing a virtual reality model of the world that is of functional use to the development and maintenance of waking consciousness". (Hobson, 2009)

And more importantly "what (Hobson) suggests is that dreaming may be a royal road to consciousness itself." Merriman (2009) describes Hobson's theory and some of its advantages in a simple way which, to me, seems excellent. It also has the advantage of an element of generality, a point I refer to later. Hobson's AIM model was already outlined in Hobson (2000).

In my opinion, this seems like a workable model - among several possible contenders - within the B series, as a more detailed frame of reference for a model of the A series within the B series. (See Yates, (2007) et al for earlier such models which could probably reasonably readily undergo appropriate modification). An important difference between Hobson's model (H, say) and such a model as suggested here (H1, say) could be that H1 is likely to contain features specifying a particular H likely to be within but not necessarily specifically interacting with a group of H1 models.

Hobson indicated that astronauts should dream much more in space than they do on Earth simply because there's more motion for them to cope with. Weightlessness takes away up and down as references. So if REM sleep promotes changes in the brain that help astronauts adapt their motor system, particularly balance, to the near absence of gravity, there is likely to be a need for more REM sleep.

Hobson and his colleagues at the Laboratory of Neurophysiology had fitted astronauts and cosmonauts with the usual "Nightcaps" (Stickgold, 1996) to record their dreams while they lived aboard the Russian space station Mir. However, NASA data collected over 6 months of flight indicated that extended space flight leads to a consistent and pronounced decrease in sleep efficiency, time spent in REM sleep, and the percent of total sleep time spent in REM sleep as measured by the Nightcap (Stickgold, 1996). As far as I am aware, that is how it still stands, whatever proferred reasons may exist.

So Hobson's theories are far from proven. Revonsuo, and others like him, claim that dreaming is not simply a random by-product of REM sleep physiology. Revonsuo claims that the form and content of dreams is not random but organized and selective: during dreaming, the brain constructs a complex model of the world in which certain types of elements, when compared to waking life, are underrepresented whereas others are over represented. All this sounds as if it could be roughly in accord with fact - which is all we need at this juncture to help model-refinement. Wargo's (2009) adversely contrarian comments on Hobson's theories do sound like reasonable folk psychology even though he says "At least Freud was on the right track. The newest theory, by J. Allen Hobson, is about as off the mark as most of the recent ideas I've read." Even bearing Wargo's comments in mind, we can come closer to Revonsuo's views and at the same time use an infrastructure somewhat like that of Hobson. Its probably not at all necessary to accept at this stage Revonsuo's evolutionary theories as this could clearly move needlessly far from Hobson.

But it may be worth pointing out that some of the views of Allan Hobson, in a similar way to those of Patricia Churchland, can present a rather dry, abstract scene, somewhat barren of humanity, which does seem rather characteristic of the intensely mathematical approach of Sir Isaac Newton and his many successors. This is a view which has been very successful in some ways, but, rather like an old plaster wall stripped of wallpaper, by its meticulous bareness and the necessary attention to often unwelcome detail, may also present us with nooks, peepholes and crannies through which we may be able to gaze on fairy seas of the soul, or on cool meadows and pastures, with a sweet flowing stream and placid cows grazing if you like... In short there may be a partly simplistic version of the A series available to us within the B series as long as we do not take there to be a precise one to one mapping of either. Whilst excessive mathematization at an early stage may not allow a comprehension or a mapping of the soul, it may nonetheless allow at least a partial representation of the soul (or of consciousness), if not in the precise terms of either A series or B series. Scarone (2009) and perhaps Sutton (2008) and Rauch (2009) may help to define the way. Clearly, too, there are great possibilities in the X-phi direction.

More may be added to the model in due course.

References

Hobson, J. Allan, Pace-Schott, E. and Stickgold, R. (2000), "Dreaming and the

Brain: Toward a Cognitive Neuroscience of Conscious States", Behavioral and Brain Sciences 23 (6): 793-842

Hobson J.A., (2009) Nat Rev Neurosci. 2009 Nov;10(11):803-13. Epub 2009 Oct 1. REM sleep and dreaming: towards a theory of protoconsciousness. . PMID: 19794431

Merriman J., (2009)
http://www.neurologyreviews.com/08%20aug/AlteredDreaming.html

McNamara P., McLaren D., Durso K., (2007), "Representation of the Self in REM and NREM Dreams", Dreaming, June ; 17(2): 113–126. doi:10.1037/1053-0797.17.2.113 ; and at http://www.ncbi.nlm.nih.gov/pmc/articles/PMC2629609/pdf/nihms53729.pdf

Rauch B., (2009), "'Natural' and Digital Virtual Realities", Leonardo Electronic Almanac, Vol 16 Issue 4 – 5 , http://www.leonardo.info/LEA/DispersiveAnatomies/DA_rauch.pdf

Scarone S., (2009) http://www.esf.org/activities/exploratory-workshops/news/ext-news-singleview/article/new-links-between-dreams-and-psychosis-could-revive-dream-therapy-in-psychiatry-585.html

Stickgold R.A., Hobson J.A., (1996), "On-line vigilance monitoring with the Nightcap", http://www.websciences.org/cftemplate/NAPS/archives/indiv.cfm? ID=19960547 or any improved version. Other interesting work using the Nightcap possibly relevant to a useful model for our present research occurs in McNamara (2007) ; NASA results referred to in the main text above are given at http://lsda.jsc.nasa.gov/scripts/experiment/exper.cfm?exp_index=846

Sutton J., (2008), "Dreaming", http://philpapers.org/rec/SUTD

Wargo E., (2009), http://thenightshirt.com/?p=115

Yates J. (2007) http://ttjohn.blogspot.com/2007_05_01_archive.html , and elsewhere in http://ttjohn.blogspot.com/

3.04 Precognition, dreams. and McTaggart's paradox

We have a very tricky problem here, as everyone realises there is a past, a present and a future and yet this fact is not expressed correctly in the existing Newtonian/Einsteinian style models of punctal (or point) time. Furthermore neuroscientists recognise the existence of the concept of the so-called 'specious present' which any thoughtful individual notices. Even that particular choice of the words 'specious present' in itself implies the real possibility of scotomisation, almost to the point of agnosia, a possibility which I have been suggesting for well over 25 years. After all, why should reality as we observe it, be necessarily specious in any way ? Possibly partly so we can simplify the mathematics, an approach which may in itself be worthy enough but is probably not a complete goal.

The common resort is that of authors like Goguen (2006), Fauconnier and many others, whose efforts even get to the level of philosophical/psychological experimental tests. Whilst obviously Goguen's work in particular, is very important in any further developments of category theory which is, itself, probably embedded in the correct approach to the problem, nonetheless in practice the method is frequently simply used for semantic collaborations within the existing ultimately Newtonian based frame of reference. Goguen has commented (dramatically and very correctly) in other papers (Goguen, 2005) that he sees no point in a speedy and unnecessary descent into what I would term mathematical metaphysics and of course I wholeheartedly agree with these sentiments !

The trouble is that an unduly semantic interpretion may savour of what might be called the opposite, you could say that it almost becomes, before it starts, a sort of 21st century logical positivist or Popperian stance of the 'metaphysics is nonsense' variety, not a good starting point that, either. I think however, Goguen's basic category theory approach is praiseworthy and worth saving. What does worry me is how few practical results of a foundational kind that we would like have, in practice have emerged from category theory, after the many thousands of pages on the mathematics.. Isham's work, for example, is clearly an exception to this, which I hope illustrates the direction of progress I would eventually hope for. And you could say that Coecke could be getting closer to the right approach, though he mainly confines his determinations to quantum theory.

What seems to be worse is the work of Penrose or Stapp, who by descending steeply to quantum theory for explanations of phenomena, is almost certainly simply wrong. (The reasons and areas of these errors were stated briefly in earlier posts.)

That being said, and the clear warning being given that we are exploring everyday phenenomena like 'past, present and future', and not simply unstructured and unproved fantasies, we continue with the blog.

It seems that apparently working cases of precognition (Note 1) often seem to have involved emotional involvement, amongst other things. From our present standpoint there are very few obviously useful studies available of precognition involving sleep which can be regarded as anything other than anecdotal, so it will be necessary to dip into the results for more conventional work in order to obtain enlightenment.

The Walker sleep lab results (Walker, 2006) seem to suggest that a 36 hour sleep deprivation period before the learning, profoundly reduced encoding of positive emotional stimuli, and to a lesser extent, emotionally neutral stimuli, but the encoding of negative stimuli was more resistant to the prior sleep deprivation.

On the other hand Tilley A. and Statham D. (1989) say that sleeping immediately prior to learning impairs subsequent retention, but did improve after 20 minutes. Grosvenor and Lack (1984) had also noted a strong detrimental effect of sleep prior to learning.

Since the latter two experiments many tests have been made of awakening sleeping people and performing tests on them. It becomes clear that the amount of sleep these people have had on previous days (Arito, 2000), whether they have been awoken during SWS, NREM sleep, REM sleep etc can give different results and interpretations. Now Patrick McNamara (2004) holds a currently minority sleep interpretation but at any rate he seems to roughly take the view that NREM sleep is used for processing the pleasant areas and REM sleep for the unpleasant areas. It seems to me that there is perhaps an as yet clearly undefined difference between emotionally negative/positive stimuli and pleasant/unpleasant stimuli (sadomasochism is only one example) as well as memories involving social situations evoking high/low anxiety levels. Obviously we try to keep it all simple and avoid hairsplitting but to note this fact may be one way to make that easier. The most commonly held belief among the scientific community (Rauchs, 2006) seems to be that REM sleep consolidates memories and aids in learning. An article in Science recently declared that "neuroscientists have long known that memory consolidation goes on during sleep." A more recent discovery is that NREM sleep may also play a role, albeit a different one, in learning. Stickgold found that different phases of sleep are tied to different types of learning. Learning visual skills, he stated, depends on the slow-wave sleep of the first quarter of the night and also on the REM sleep of the last quarter. Learning movements relies much more on the NREM in the later part of the night.

In 2001, Wilson found that rats dream about their activities (i.e. running through a maze) during NREM sleep in addition to in REM sleep. Unlike during REM replay, where the experience occurs approximately in real time, the memory segments that were replayed during NREM seemed to be snippets of experience. Also, unlike REM sleep, slow wave sleep seemed to replay only what had happened immediately before and not something twenty-four hours ago. Because of the possible time-delay REM memory reactivation, it might be representative of a more gradual reevaluation of slightly older memories.

From the Walker (2006) sleep labs I further quote: "Within the sleep control group, both positive and negative stimuli were associated with superior retention levels relative the neutral condition, consonant with the notion that emotion facilitates memory encoding. However, there was severe disruption of encoding and hence later memory retention deficit for neutral and especially positive emotional memory in the sleep-deprived group, which exhibited a significant 59% retention deficit relative to the control condition for positive emotional words. Most interesting, however, was the resistance of negative emotional memory to sleep deprivation, showing a markedly smaller (19%) and nonsignificant impairment.

Taken as a whole, these studies suggest a rich and multifaceted role for sleep in the processing of human declarative memories. Although contradictory evidence is found for a role in the processing of simple, emotion-free declarative memories, such as the learning of unrelated word pairs, a substantial body of evidence indicates that both SWS and REM sleep contribute to the consolidation of complex, emotionally salient declarative memories, embedded in networks of previously existing associative memories. In light of this evidence, pronouncements of a lack of relationship between REM sleep and "memory" appear to be unfortunate overgeneralizations that

91

disregard evidence that specific sleep stages play distinct roles in different stages of memory"

When a tranquilizing drug was given to rats 5 min after rats were put in a "fear chamber"(Bustos, 2006) a second time, after they had learned to fear impending foot shock (even though feet were never shocked again). When retested some 10 days later, the rats that had been given tranquilizer during the earlier recall trial showed little freeze behavior, indicating that they had forgotten what they learned and what they were forced to rehearse when they were put into the fear chamber a second time. The drug treatment and its timing caused them to forget to be afraid. Note that there was a rather surprising finding that the drug had no convincing impairing effect when given during the consolidation period of the initial learning trial. But when given during a re-consolidation period of a second rehearsal trial, profound forgetting effects were noted. The drug interference effect could be seen for up to 60 minutes after re-consolidation, not later.

This suggests strongly that during recall, a memory trace becomes vulnerable once again. The memory may be lost altogether. The memory can also be altered, leading to a false memory. On the other hand, the memory can actually become enriched, if re-consolidation involves new information that expands the amount of information stored and the improves the quality of the original information. So, remember, that even though rehearsal promotes retention, it may not always be helpful. Many advise being careful about what happens during the process of recall, staying focused during rehearsal of things you are trying to remember, and making certain there are not distractions or extraneous information being inserted into your memorizing process.

Frank Logan's NMU handbook mentions "One of the memory mysteries is the prior sleep effect. It has long been known that sleeping after learning is beneficial to memory. This may be because learning is consolidated during sleep or it may simply be that nothing happens during sleep to interfere with what was learned. Whichever, we may ask whether sleep before learning also affects memory. It does, but the effect is negative! When sleeping people are awakened and given new material to learn before going back to sleep, they seem to learn perfectly well. But when they are later tested over the material, they remember very little. How much they remember depends on how long after waking before the learning occurs. There is some negative prior sleep effect for up to an hour. This may be one reason that dreams are quickly forgotten. One implication of the prior sleep effect is that you best not schedule study time immediately after sleeping or napping. A second implication is that, if you have an early morning class, you should be sure to get up early. A final implication is that, if you doze off during a lecture or while studying, you not only lose that time, you won't remember much of what happens next. Students need plenty of sleep, but you should be sure to sleep after, not before learning."

At this point we may well be best considering Harnad's symbol grounding problem, and the interpretations of Barsalou in 1999 and the Glenberg and Kaschak (1997, 2002) studies. In this case here, at its simplest, Harnad's problem is that in order to be able to interpret and use a symbol, you'll need another symbol that expresses its meaning, and so on. Glenberg and Kashak try to avoid this in a series of experiments relating reaction time to actual movement. Participants read sentences one at a time, some of which made sense and some of which did not. They had to press one of three buttons, one close to the body, one about arms length from the body, and one in between the two.

When participants read the sentence "Open the drawer," for example, they will respond that it makes sense faster when their answering involves moving from the middle to the closest button.

But, our own studies to date suggests that there won't be a simple optimality model along the lines of that of John McNamara.(2001) i.e. a semi-markovian model or even one using a fairly easy chaos (or biological statistical) interpretation. John McNamara was certainly 'sold' on the idea of using category theory in his later years but this, too, has to be done very basically as I've tried to do earlier. And there is no need yet to use an image-based or action-based system yet – necessarily.

We also want to avoid so far unresolved logic-based problems (as in the Wason selection test) as far as we can. We will conclude so far that further experiments along the lines of our earlier post at http://ttjohn.blogspot.com/2006/04/do-we-dream-of-future.html may be possible but not easy in concept. Some incorporation of ideas referred to in Levy (1996) but modified for present purposes may help.

References

Arito, H., and Takahashi, M. (2000) "Maintenance of alertness and performance by a brief nap after lunch under prior sleep deficit," Sleep, 23:813-819 .

Barsalou L.W., (1999) Behavioral and Brain Sciences 22, 577–660

Bustos, S. G, H. Maldonado, and Molina, V. A. (2006), "Midazolam disrupts fear memory reconsolidation". Neuroscience. 138: 831-842.

Butz M.V., Sigaud O., Gérard P, (2004) , "Anticipatory Behavior: Exploiting Knowledge About the Future to Improve Current Behavior" in "Anticipatory Behavior in Adaptive Learning Systems: Foundations, Theories, and Systems" Springer Berlin / Heidelberg , ISSN: 0302-9743

Glenberg, A. M. (1997). What memory is for. Behavioral & Brain Sciences, 20, 1-55.

Glenberg, A. M., & Kaschak, M. P. (2002). Grounding language in action. Psychonomic Bulletin and Review, 9, 558-565.

Goguen J., (2006) "Mathematical Models of Cognitive Space and Time", (Newly Re-Revised, 1 May 06), a preliminary version will appear in 'Reasoning and Cognition', edited by Daniel Andler and Mitsu Okada; proceedings of symposium at Keio University, December 2005 ; also at http://cseclassic.ucsd.edu/~goguen/pps/taspm.pdf

Goguen J., (2005), "Ontology, Society, and Ontotheology", wwwcse. ucsd.edu/~goguen/pps/fois04.pdf For example, where he says ""In Western culture, mathematical formalisms are often given a status beyond what they deserve."

Grosvenor A, Lack L.C., (1984), Sleep. 7(2):155-67.

Holley, J., Pipe, A.G. & Carse, B. (2004) Dream Function as an Anticipatory Learning Mechanism. In M. Butz et al. (eds) Anticipatory Behaviour in Adaptive Learning Systems: Proceedings of the 2nd ABiALS Workshop, pp31-40.

Levy B., (1996) "Improving memory in old age through implicit self-stereotyping", J. Personality and Social Psychology. 71 (6): 1092-1107.

McNamara J. M., Houston A.I., Collins E.J., (2001) "Optimality Models in Behavioral Biology" SIAM REVIEW, Society for Industrial and Applied Mathematics Vol.43,No.3,pp.413 –466

McNamara P., (2004), "Evolutionary Psychology of Sleep" (Praeger)

Rauchs G., et al (2006), "Sleep after spatial learning promotes covert reorganization of brain activity", Proc Natl Acad Sci U S A 2006 Apr 24;[epub ahead of print].

Tilley A, Statham D. (1989) Acta Psychol (Amst). 1989 Mar;70(2):199-203.

Walker M.P., Stickgold R, (2006) "Sleep,Memory, and Plasticity", Annu Rev Psychol 2006; 57:139-166

Watzlawick P., (1983), "The Situation Is Hopeless, But Not Serious", W. W. Norton & Co.

Notes

1. By 'precognition' in this blog post I simply mean instantiations of the realization of the future by some means, which could be conjecture, hypothesis, predictions, mathematical predictions, mental pictures, dreams, revelations or some other as yet specifically undefined way. It is something which in essence has to be real and solid. We know that we can predict tomorrow and even visualise it. It may be that to actively foretall it as a soothsayer or fortuneteller etc might claim to do successfully, could fall into a rather extreme bracket, but that in itself is a matter of opinion, and following say Mercea Elaide, could even be a matter of social context and spiritual comitment to it.

The thing we should hold to is the idea that we are dealing with something which is real and observable everywhere, in fact it is part of the way our minds work. We do not wish to be unduly judgemental about what its specific manifestations are (or should be) at a very early stage in the piece. To do so would be totally wrong and misleading, perhaps as wrong in its own way as to fall into the trap which Goguen (2005) warns us of.

Simple attempts to fathom anticipatory behaviour are referred to for example in Butz (2004) and experiments described in Holley (2004).. We could consider Watzlawick (1983) and his views on the past as they could also relate indirectly to the future, Pattee, and von Uexkull and his ideas of achieving of common realities, Cariani etc. and whilst we bear them in mind, this blog is for the moment taking a somewhat differing viewpoint.

Time Travel in Theory and Practice

4

Experimental Philosophy and other issues which may help the MBI

Experimental philosophy (or X-phi) is discussed in this chapter, relative to the other topics in this book.

Section 4.01 is a general and non comprehensive introduction to X-phi, partly in historical terms, and then X-phi is related to the MBI or 'Many Bubble Interpretation' and to the merological fallacy, the multiple universes of Max Tegmark and others, the MWI or 'Many World Interpretation' of Everett and later workers, teletransportation and brain cloning (somewhat in the manner of Parfit), and then the use of the new MBI interpretation for dreamwork, global workspace theory and the like is discussed.

Section 4.02 describes how our experimental findings using X-phi support the Many Bubble Interpretation.

Section 4.03 discusses how the use of X-phi may vary between cultures, its general methods of operation, and then we proceed to relate it to the Many Bubble Interpretation, to the so-called 'Libet half-second', to the work of Ariely, and to the work of Hacker, Ed Vul and others.

4.01 Do Intuitions about Reference Really Vary across Cultures?

Abstract

We discuss whether intuitions about reference really vary across cultures and how these variations relate ultimately to the McTaggart A-series. We conclude that much more work needs to be done, and suggest how it can.

Introduction

In the present note we seek to establish and promote new results, in some cases using experimental philosophy (a subject which we have considered for many years (Yates, 2008b)) where it becomes necessary. These new results at present usually relate to the McTaggart A and B series and to the study of time (Yates, 2008, 2008a). So the aim is not to criticise existing X-phi results - I am truly pleased and glad that the field is obtaining a foothold - but in order to obtain practical results it is necessary to point out when or where more work needs to be done, to establish usable answers to existing problems, whilst still keeping research not too lengthy.

Details

Questions like the one of the current heading query above (Machery, 2009) have been repeatedly raised. Of course they go right back to Mill, Kripke, and more recently Machery (2004), Sytsma (2009), Lam (2009), and now yet again Machery (2009).

Of course, others, in particular Frances (1998) and Sosa (1996), have worked to resolve Kripke's puzzle. In fact on the face of it, the exposition of Frances (1998) on Millian theories sounds fine to me, up to a point. For present purposes, perhaps in the exercise in Machery (2004), there are at least two issues, as many philosophers might say. One is whether or not the name

"Godel" in Kripke's fictional scenario has to have the same meaning as the actual name. The second issue is whether acceptance of the coherence of the fictional scenario already commits us to Millianism.

At this point we need to consider Machery (2004) which plainly states that two views, the descriptivist view of reference and the causal-historical view of reference, have dominated the field. In any case, certainly the work of Machery et al becomes important if we are to consider the latter classification. At this very point the power of X-phi arises, whether or not Machery's eventual conclussions are correct.

In this connection Sytsma (2009) points out an objection raised by Sosa (2007). In fact in footnote 5 of Sytsma (2009), Systma points out he "considers that "defense against experimentalist objections to armchair intuitions is anchored in the fact that verbal disagreement need not be substantive". In this context, if the results of Machery (2004) reflect divergent interpretations of the probe, then it is not clear that the variability shown reflects differences in the semantic intuitions at issue for the philosophical debate. One issue such an objection raises is how to decide where the burden of proof lies. Sosa continues: "The experimentalists have, so as to show that supposedly commonsense intuitive belief is really not as widely shared as philosophers have assumed it to be. Nor has it been shown beyond reasonable doubt that there really are philosophically important disagreements rooted in cultural or socio-economic differences". Although we cannot argue the point here ......". Sytsma thus admits that they are not arguing with Sosa but goes on to claim roughly that Sosa is seeking too high standards of proof. Now I would say the problem may be more that X-phi practitioners need to actually reach believable standards of proof with economic amounts of data.

The problems with Kripke have frequently been discussed by the aforenamed experimental philosophers. (Machery (2004, 2009), Sytsma (2009), Lam (2009)) actually seem to be in essence doing armchair work, once the formality of doing brief surveys has been met. In short, the armchairs remain (at least partly) unburnt !

To give a related example of what I am saying, Knobe et al (2009) says of compatibilism and incompatibilism "In our view, the data presently available is not sufficient to decide between these contrasting hypotheses. In short, there is still much work to be done. And while the problem of free will has historically been the prerogative of philosophers, the current study suggests that researchers everywhere who investigate folk psychology, folk physics, and moral cognition have contributions to make in solving this particular puzzle". Now Knobe's work was carried out in United States, Hong Kong, India and Colombia and the authors still have that view. As far as I know only domesticated American and Hong Kong cases were dealt with by Sytsma and Cantonese diaspora cases by Lam and Machery. And Hong Kong is compact, developed, relatively modernised (with a better modern skyline than Manhattan) and not typically Chinese as much of mainland China is.

Of all the above cases, diaspora cases do not sound the best cases to use to attack Kripke's argumentation, as the cultural references presumably refer partly and possibly primarily to the host country, normally the USA for these diaspora studies. It is all very well to effectively go to the local fish and chip shop or deli to make your foreign language queries and in fact Knobe's early work (in English) was done by asking questions in Central Park, NY., and this is a very legitimate way to get a general local feeling, but world anthropology and evolutionary psychology and its conclusions at Tooby and Cosmides level really are another matter. It is far better and often essential to go back to source. At the "Institute for Fundamental Studies" (which at present has main headquarters in UK, Maharashtra (India), and Goa) we normally deal with non-diaspora Hindi, Marathi, Konkani, Tamil and English speaking cases, and we get gratifying results.

Machery (2009) says "So, what's going on?". Well, the above is some of what's going on.

But there is much much more !

I am saying that in a further three ways at least - and simply as a beginning - that more care must be taken.

Firstly I mention Vul (2009) and Haynes (2008). On comments on Haynes' work on free will, for example, Auburn University Professor Roderick T. Long (2009) says "This is a hopelessly bad argument; the results of this study have nothing to do with the free will issue at all. This is simply a case of experts in one field (neurophysiology) thinking they are experts in another field (philosophy) that they seem to know very little about." To be fair, Haynes himself did start his career briefly in philosophy but most free will philosophers either ignore Haynes's work, or deny free will already, or are seeking a work round.

Fortunately I do not seem to need a work round as Haynes's work seems to

100

provide simply more evidence that the McTaggart B series is insufficient and we need the A series as well. Perhaps more details later (Yates, 2008, 2008a, 2008b, 2009).

But my opinion aside, Long's general position (though not necessarily his views on politics or economics) is quite widely held. Professor Colin Blakemore, a neuroscientist and director of the Medical Research Council, apparently said (Guardian, 2009) : "We shouldn't go overboard about the power of these techniques at the moment". I certainly agree ! It seems that X-phi has still largely to come to terms with Haynes' work, but the eager assumption of a very simple interpretation of results such as those of Haynes, should certainly not be made. I refer to particularly to the recent work of Vul (2009) concerning MRI interpretations and also to the implications of the work of Hacker (2003) but detailed discussions on both could add considerable additional material to the above.

Secondly I mention the important work that has recently been done in behavioural economics, and in particular the work of Ariely (2009). This helps to bring yet more clarity to the view that the old idea that market approach which presumes that "the common people know what they want" is actually quite wrong. Ariely (2009), who is Alfred P. Sloan Professor of Behavioural Economics at M.I.T. has written many papers to this effect. This work undoubtedly affects questionnaire design and we all need to consider these angles. This sort of matter goes well beyond minor details of presentation. Most Westerners do not know or care, for example that the colour "white" is a colour for weddings and the like in the West, but anyone who lives in India can hardly miss that in India, "white" is the colour for funerals and "red" is the colour for weddings ! But Ariely's work, which is not per se given cross-cultural connotations in his experiments, must have its conclusions considered in such ways in each and every local context - when we are considering reinterpretations of philosophers like Kripke.

Finally we have prejudice.... Unfortunately it does not begin and end with Engine Charlie Wilson's dictum "What's good for General Motors is good for the USA". The Implicit Association Test (Nosek, 2009) has its most surprising and controversial finding as its indication that about 70 percent of those who took a version of the test that measures racial attitudes have an unconscious, or implicit, preference for white people compared to blacks. This contrasts with figures generally under 20 percent for self report, or survey, measures of race bias. Current studies in the research came from a number of countries including Germany, the Netherlands, Italy, the United Kingdom, Australia,

Canada, Poland and the United States. They looked at such topics as attitudes of undecided voters one-month prior to an Italian election; treatment recommendations by physicians for black and white heart attack victims; and reactions to spiders before and after treatment for arachnophobia, or spider phobia.

Obviously the IAT does not apply to white people only. One might apply it to Iranians living in Iran, for example, and their views on non-Iranians. Most certainly it will influence all who give or take such tests, to a greater or lesser degree. I think it was Joshua Knobe who did somewhat similar tests on philosophers as compared to lay people, and found such a bias there, but in his tests it acted as a reverse bias. At this point we could well become worried about relativism and hermeneutics in the sense of Heidegger and Gadamer.

A further brief point I'd like to make is the question as to whether this approach to Kripke involves modern semiotics quite directly rather than simply semantics. I hesitate to mention Barthes, Saussure, Lacan and so on but their conceivable relevance seems obvious. David Sless (1986) remarks, 'semiotics is far too important an enterprise to be left to semioticians' and it may well be true.

Conclusion:

This note is not to be in denial of progress, just to say progress may be difficult and when back at the "Institute for Fundamental Studies" in Mumbai after the monsoon I intend to do some investigations myself, bearing in mind the earlier work of Kripke, parallel universe ideas like those of Deutsch, Parfit etc., and the approach of Noe and of Clark and Chalmers. Naturally all this may ultimately give further evidence for the Many Bubble Interpretation, involving the A series of McTaggart.

References

Ariely D., (2009), http://web.mit.edu/ariely/www/MIT/papers.shtml ; e.g. "Tom Sawyer and the construction of value", Journal of Economic Behavior & Organization, Vol. 60 (2006) 1–10 ; popular book: "Predictably Irrational", Harper Collins, (2008), etc.

Frances B., (1998), Mind 107, 703-727.

Haynes J-D., (2008),
http://medgadget.com/archives/2008/04/not_a_free_will_after_all.html

Hacker P., Bennett M., (2003), "Philosophical Foundations of Neuroscience", Wiley-Blackwell, ISBN-10: 140510838X, ISBN-13: 978-1405108386

Lam B., (2009), http://philpapers.org/rec/LAMACS ;
http://faculty.vassar.edu/balam/arecantonesespeakersreallydescriptivists.pdf

Long R.T., (2008),
http://medgadget.com/archives/2008/04/not_a_free_will_after_all.html

Knobe, J., Sarkissian, H., Chatterjee, A., De Brigard, F., Nichols, S. & Sirker, S. (forthcoming). Is Belief in Free Will a Cultural Universal? Mind & Language ; http://www.unc.edu/~knobe/cultural-universal.pdf

Machery, E., Mallon, R., Nichols, S., & Stich, S. (2004). Semantics, Cross-cultural Style. Cognition, 92, B1–B12.

Machery E., (2009),
http://experimentalphilosophy.typepad.com/experimental_philosophy/2009/06/do-intuitions-about-reference-really-vary-across-cultures.html

Nosek, B. A., Smyth, F. L., Sriram, N., Lindner, N. M., Devos, T., Ayala, A., Bar-Anan, Y., Bergh, R., Cai, H., Gonsalkorale, K., Kesebir, S., Maliszewski, N., Neto, F., Olli, E., Park, J., Schnabel, K., Shiomura, K., Tulbure, B., Wiers, R. W., Somogyi, M., Akrami, N., Ekehammar, B., Vianello, M., Banaji, M. R., & Greenwald, A. G., (2009), "National differences in gender-science stereotypes predict national sex differences in science and math achievement", PNAS published online before print June 22, 2009, doi:10.1073/pnas.0809921106; Greenwald
https://implicit.harvard.edu/implicit/research/ ; Sriram N, Greenwald A.G., (2009), "The brief implicit association test", Exp Psychol. 2009;56(4):283-94 ; http://faculty.washington.edu/agg/pdf/BriefIAT.26Jan09.pdf

Sless, D. (1986), In Search of Semiotics. London: Croom Helm

Sosa, D. (1996), "The Import of the Puzzle About Belief," The Philosophical Review, 105, 373-402.

Sosa, E. (2007), "Experimental Philosophy and Philosophical Intuition",

Philosophical Studies, 132, 99–107.

Sytsma, Justin and Livengood, Jonathan (2009) A New Perspective concerning Experiments on Semantic Intuitions. In [2009] Society for Philosophy and Psychology, 35th Annual Meeting (Bloomington, IN; June 12-14).

The Guardian, Friday 9 February 2007, Colin Blakemore as quoted therein: http://www.guardian.co.uk/science/2007/feb/09/neuroscience.ethicsofscience

Vul E., Harris C., Winkielman P., Pashler H., (2009). Voodoo Correlations in Social Neuroscience. Perspectives on Psychological Science, in press ; Vul E, Kanwisher N. (in press). "Begging the question: The non-independence error in fMRI data analysis". To appear in Hanson, S. & Bunzl, M (Eds.), Foundations and Philosophy for Neuroimaging.

Yates, J. (2008a). http://cogprints.org/6176/ , "Category theory applied to a radically new but logically essential description of time and space", PHILICA.COM, Article number 135.

Yates, J. (2008b), http://cogprints.org/6232/ , "Experimental philosophy and the MBI", PHILICA.COM, Article number 139.

Yates, J. (2008), "A study of attempts at precognition, particularly in dreams, using some of the methods of experimental philosophy." , Philica.com , Article number 146.

Yates, J. (2009), "The Many Bubble Interpretation, externalism, the extended mind of David Chalmers and Andy Clark, and the work of Alva Noe in connection with Experimental Philosophy and Dreamwork", http://ttjohn.blogspot.com/2009/05/many-bubble-interpretation-externalism.htmlA

4.02 A study of attempts at precognition, particularly in dreams, using some of the methods of experimental philosophy.

Abstract:

Actual situations where folk philosophy might have predicted precognition effects were studied and dealt with experimentally and theoretically. Extremely strong experimental results were obtained but the findings supported not precognition but the Many Bubble Interpretation, which uses at this time dynamical systems theory as applied to the physics of the brain. Further experiments and theoretical work were discussed.

Introduction:

Blackmore's (2002) analyses are possibly the most up to date detailed appropriate account of controlled trials on precognition. Her remarks go back as far as the early dream work of Lord Kilbracken. Any results reported in Blackmore (2002) or implied by it seem to suggest that the subject does not present much future hope for precognition. Specifically with regard to dreams, Hobson (2005, 2006) doubted if there is any precognitive element in dreams though he seems to have had at least one dream which could be fitted to that category (a totally different thing, of course). I am largely in agreement with some aspects of Hobson's position on interpretations to date though there is still much exciting work to do, some of which I begin in this essay .

I have already considered the dreamwork of Domhoff (2002, 2003), Hobson, Metzinger (2004) and others elsewhere (Yates, 2008). The point has to be made that Domhoff has tried to computerise many aspects of dreamwork and his approaches to the multifarious problems of detail and interpretation plus his genuine attempts to involve internet interactions have to be borne in mind at all times, if not necessarily to be followed. Hopefully further experiments may be carried out at least partly on the internet, but many additional considerations, including those of experimental philosophy, will need to be dealt with.

Generally speaking, Hobson's and Blackmore's results and the largest fraction of similar work carried out with reasonable scepticism suggests that precognition does not happen, inside or outside of dreaming, under the constraints and conditions which have been imposed to date. I think that both Hobson (2006) and Blackmore (2005) were willing to be convinced otherwise by an effective proof, and therein lies the rub.

However we have already pointed out in earlier essays (Yates, 2008, 2008a, 2008b) that existing statistical methods may be inappropriate here and also that there is much more to be found or interpreted through observations.

I have also looked at other statistical and quasi-statistical aspects of the situation and have come to the conclusion that the methods of experimental philosophy have led to surprisingly exciting insights into dreamwork from a rather different angle.

Of course I refer to the continuing work of Schwitzgebel (2002, 2003, 2006, 2009) into colours in dreams as being seminal in this regard, and also with regard to specific methodologies.

Accordingly and also consequent to Yates (2008a,2008b), some experiments have been carried out on a number of subjects, as detailed below.

Whilst we have said that conventional statistics will not give a full picture of events (Yates, 2008) we now point out relatively conventional alternative methods by which dream studies can be carried out (Schwitzgebel, 2002, 2003, 2006, 2009) and there is doubtless much more to come. Schwitzgebel (2005) favours up to a point the classical traditional methods of the Titchener school of psychology and we could see that these can be leavened by some comments of Sosa (2007) (for example "if philosophers are ill-equipped to probe the brain in the ways of neuroscientists, it would be easy enough to broaden the movement's self-conception to include interdisciplinary work, provided neuroscientists care enough about such issues with philosophical import, as no doubt some already do. Indeed, many experimental philosophers would probably define the movement in this interdisciplinary way").

We also have to bear in mind that much work of the Titchener school seems to have been only repeatable precisely enough by the Titchener school. I refer particularly to the chequered history of the Perky effect (Segal , 1964), (Martens , 2005) as a clear instance, but have to point out that Baars (2003), for example, still seems to take the Perky effect, or what it seems to imply, quite seriously and we have to consider the at least roughly feasible interpretations of Brockmole (2002). So whilst we cannot but concede that the general Titchener approach can be taken as somewhat of a curate's egg (addled but good in parts, as the curate proverbially says), it is a valiant attempt at a difficult problem in consciousness and as Schwitzgebel (2005) says, can be suitable for improvement. Schwitzgebel's own work at least hints as to directions which this may take.

The Experiment as Performed

This experiment helps us to discover the degree if any of backward leakage using the present approach which can be obtained by dream studies. The MBI ('Many Bubble Interpretation') approach and the importance of McTaggart's work certainly do not stand or fall on the basis of such studies but they could provide a pleasant confirmation of its correctness. The MBI may ultimately replace the present crude 19th/20th Century idea of punctal time, which seems like only a dim shadow of reality in comparison.

In Stickgold's experiment he had his subjects perform sequences of tasks and then showed that many of the subjects subsequently dreamt about these tasks. The present experiments record the dreams and then have the subjects carry out the tasks, which of course are chosen prior to the dreaming without informing the subjects in any way of the tasks, prior to the dreams. In terms of folk psychology, successful results could be regarded as precognition but according to the Many Bubble Interpretation, it is simply backwards leakage.

One idea will be to then try to obtain some parameters to enable us to improve our existing dynamical systems models (Yates, 2008) along the lines of the work of Hannon and Ruth (1997) or using other mathematical brain models such as those of Baars, Franklin or Koch, appropriately modified.

But first we will describe briefly the original Stickgold experiment. Following this we give an explicit mathematical account of how a reverse Stickgold effect can be produced. Then we will go on to obtain results which may help to confirm or indicate the reverse Stickgold effect, referred to in an earlier paper.

Stickgold's experiment

Stickgold (2000, 2003, 2005) controlled the content of 17 different people's dreams after the first hour of sleep. Twenty-seven test subjects played Tetris on Nintendo sets for three days, with a two-hour morning session and a one-hour evening session the first day, and a one-hour morning and evening session the following days. Of the 27 people, 12 were beginners to the game and 10 were experts. Five of them were amnesiacs as well. Seventeen members of the group recalled dreaming of falling Tetris pieces at least one hour after falling asleep. Most of the dreams occurred the second night.

As far as I am aware these are the first major trials in history which so consistently seem to induce specific and definite dreams, and as such they should be extremely relevant to the present backward leakage study. The problem with psychological tests as compared to simple physics measurements has usually been the great difficulty in obtaining clear and consistent results. The chequered history of, for example, the Perky effect which I mentioned in detail earlier is an extreme but only too typical example. At the other end of the scale we have, say, Milgram's (1974) torture experiments which seem to have had very high repeatability all over the world.

Mathematical Representation of Stickgold and reverse Stickgold effects

To write down a mathematical model of the reverse Stickgold effect (Yates, 2008, 2008b), we use experimental philosophy and the work of Pizarro (2006).

This concerns the ripple effect. Now we consider a simple B1 series representation of the A series bubble. This is only a preliminary model and is not necessarily an accurate description. Because the A series cannot be precisely mapped onto the B series the model will never be completely accurate, because it cannot be. Thus the B1 series representation may appear to have substantial weaknesses as compared to a bona fide B series, and indeed may be inconsistent with it. We regard a present time bubble as PaPrFu(n) at Tn.

The ripple effect - the giving of an individual at a present time, information about an event which he remembers, is supposed according to Pizarro (2006) to alter his real (perceived) memory of the event. Here we have a physical effect on the brain from an applied information input. Say by an application of information f in the present the new configuration becomes PafPrfFu?(n) . Paf is the modified memory. Prf is the new present situation. We have not filled in Fu? because we do not need to for present purposes.

Now we need to remember that we are dealing with the neurology of the brain and presumably other factors. Now along the B or the B1 series we have a simple timeline. So one way of writing it would be that at T(n) we have one bubble PafPrfFu?(n) and at T(n-) , an earlier time, we have PaPrfFuf(n-). This is only a rough preliminary model but we are talking about real neural and perhaps other configurations at two different times. I do not deal with Fu?(n) at T(n) because in the present treatment we do not need to, but the work of Hohwy and Frith (2004) and others, makes it clear that Fu? is likely to be a real 'physical' configuration in a real 'physical' B series.

Here Fuf(n-) represents the future in the B1 series where there is going to be a a perturbation f. At this point we do not have to immediately consider many of the problems and paradoxes which one might normally expect as we know that it is not physically likely to be possible to get a completely accurate and consistent B1 series model as we would expect in the B series.

Now what is the relation of PafPrfFu?(n) to PaPrfFuf(n-) ?

Well these are two representations of a individual at times T(n) and T(n-). They are both brain models which can or should be each writeable down consistently and mathematically although we are not certain of their mutual consistency.

Well we specifically defined PafPrfFu?(n) above. As for PaPrfFuf(n-) , common sense hopefully tells us that it will exist too. It seems to mean that could well be a future perturbation of Fuf(n-) and that this will also occur in a simulated present Prf(n-) which of course is combined with past and future to form the bubble PaPrfFuf(n-) at T(n-).

So at T(n-) we should be able to write down, or to predict, the future. It is possible that it will only be a weak prediction, at the present state of the art, as we remember that Pa(n) was clearly weak.

But the general point is made that, in the B1 series at least, we can perturb the system by a perturbation f at time T(n) and that this will appear in the bubble PaPrfFuf(n-) as Fuf(n-). One's first reaction is that in a normal block time B series, this would be expected as that is how mathematics works. We might take the view that it could hardly be any other way anyway, and that toy models should also wrap up quantum and chaos effects in the same kind of system.

But this is a B1 series in the MBI . And the existence of the bubble PaPrfFuf(n-) implies that at time T(n-) we already have in our system enough information to write down (or if you like in folk philosophy terms 'forsee') something concerning the future. And we can use methods like those, for example, of Hannon and Ruth (1997) to actually mathematically represent the system at time T(n-) and to include the future of that system at T(n).

So now we have to ask what PaPrfFuf(n-) is in real terms. Well one representation might be a person in a particular psychological state. For example the dream state in Stickgold's experiment seems to represent Paf2Prf2Fu?(n) where f2 is a perturbation in the past (in Stickgold's case, the playing of Tetris), mirrored in a present dream state Prf2(n) (where Tetris is presumably dreamt about), in a simple situation where future involvement is not concerned.

So the equivalent representation of PaPrfFuf(n-) is the dream of a future perturbation, perhaps the playing of a game of Tetris in the future. All this does not prove the matter but it makes it clear that it can be written down mathematically using say the methods of Hannon and Ruth (1997). So perhaps the big question is: how reliable and consistent can such experiments be made ?

The work of Jones and Pashler (2007) suggests that prediction is never superior to retrodiction, even when subjects are forewarned of a forward-directional test. Only 217 and 353 subjects were used in their two experiments and of course the test was carried out subsequent to all the images to recall being memorised. It has been suggested that prediction may be an organizing principle of the mind and/or the neocortex, with cognitive machinery specifically engineered to detect forward-looking temporal relationships, rather than merely associating temporally contiguous events. There are not many tests for this idea, other than Jones (2007). The fact that Jones's work seemed to show no evidence of temporal asymmetry tends to bode well for the more advanced cases we consider in the present paper.

Factors taken into account in experiment planning. These include the idea suggested by Montague, Hyman, and Cohen (2004), it may be that events as reward or punishment cause prediction-focussed mechanisms to become active, whereas affect events like those used in Jones (2007) do not. The common observation that people are better at reciting the alphabet forward than backward also reflects the existence of inherently directional motor plans. A temporal asymmetry confined to sequential motor plans that have been repeatedly performed is quite different from an overall specialization of the memory system for prediction, however, although it could point to prediction improvement methods.

Freewill, intentionality (Malle (2001, 2004) and such ideas as free will illusionism (Nadelhoffer , Feltz , (2007) may be the subject matter of further discussion and also they could have an obvious role in future experiment design.

Reverse Stickgold effect experiments

We did several studies. Here are two of them:

(1) This was essentially a repetition of the Stickgold experiment using 8 of our own subjects but recording dreams on evenings 7 days before and 7 days after, the Tetris plays.

The dreams were recorded by the subjects themselves, in English. The subjects also filled in a brief questionnaire as to dreaming habits. The subjects were from 6 to 15 years of age. Their mother tongues were Marathi and Hindi, but all could write and speak fluent English, as they were students at a local school where English was taught as a first language. A primary reason for their learning of English was allow personal advancement in whatever sphere of life they were later to lead. Their level of English presentation and expression was thus of a higher standard than would be the case in the average UK school for children of the age range.

Subjects were not told why or how the experiment was being carried out. The Tetris console was simply supplied to them in the middle of the testing session, when they were preoccupied with school, their hobbies and other such things. The subjects had just been asked to record dreams for a fortnight and, in mid session, were given the use of a Tetris gameboard and told to play Tetris a lot over a brief mid session period.

Briefly, the results were that for the 8 subjects there were 10 dreams of a probable Tetris type before play and 6 such dreams after play. Of these dreams, there were 4 very Tetris like dreams before play and none after play. These dreams averaged over the subjects, and there were no notable peak scores. One subject reported no dreams at all after play and one reported no dreams during the entire session.

(2) In this experiment, another group of subjects gave detailed answers to a dream questionnaire. Incidentally, the contention of Schwitzgebel (2006, 2009) that few people dream only in black and white was true for all our subjects. No test subject said that they dreamt solely in black and white. We used 13 subjects, aged between 8 and 15. The mother tongue of all these

subjects was Marathi, but they all spoke good Hindi and reasonably fluent English as well. Detailed dream results were collected for four days before, and four days after the mid session day. Details were taken down carefully by an experienced and quadrilingual test assessor over a period of over 10 days and for many hours per subject.

I do not at this point propose to give a quantitative assessment of the results but I bear in mind Stickgold's apparent contention that form rather than specific substance is what is best measured. i.e. in his Tetris and Alpine Skier experiments he was looking generally in the first case for activities which had the same qualities as Tetris and Alpine Skier. Thus in the case of Tetris he might have considered simply pieces moving in the air or at a pinch even raindrops or such like of an appropriate design or pattern. In the case of Alpine Skier he seems to have been looking for the visceral effect of someone actually skiing, or perhaps one of the more advanced multicolored Virtual Reality ski or switchback games now to be found at the better grade of amusement parks. Specifically I saw an excellent such VR game at the seaside in Blackpool, England some years ago. Such games are roughly like the switchback equivalent of a Link Trainer for pilots, and ambitious home construction details are available on many websites (Wikipedia, 2008)   There is clearly scope for more ambitious experiments in this regard but the time and effort involved will mean that careful advance experiment planning, relatively speaking as detailed, thought out, and meticulous as the Titchener school had expected to realise, is likely to be required.

At the midsession period, Tetris was not played but a small gift given to each child, of a kind they might like. Examples are a remote control toy car and a remote control toy helicopter. Scoring was based on dreams about the chosen object, i.e. car, helicopter and so on. They were not told before the test what they were getting as a gift. This was a relatively poor area and the subjects rarely received gifts, in fact some had never had any gifts before.

Be all that as it may, on the broad criterion above, which counted some cases of running in motor traffic as amounting to a dream of a car, and so on, the score was 10 subjects having prior (in folk philosophy perhaps precognitive) dreams and 9 having subsequent dreams. One subject claimed to have had only one dream during the entire period and one subject could only recall a vague prior dream. Now this is an extremely good prior score !

On a more narrow criterion, where only dreams specifically about an object exactly the same as the chosen object were included, the prior score was again high, being 6 prior with 7 subsequent dreams. The total number of all dreams per subject recorded was not high, being on average 2 or 3 before and 1 or 2 subsequently. The criterion here was that if they got a toy car (for example), a positive result would be if they dreamt about a car. One subject's prior dream was not only about a car, but he specifically dreamt the correct color which he mentioned without prompting, being the only subject to mention the relevant color (yellow) of the significant object in the dream series.

Now these subjects were all impartially and individually quizzed and questioned over a period of time, without any leading of them in a particular direction. But of course we cannot yet draw too many conclusions, nor determine whether we will always get the same result. One interesting problem (Haidt , 1993) concerns the idea of quizzing subjects about their views on a story in which a person has a pet dog, which unexpectedly dies, probably by accident. The dog is then eaten by its owner. Most people react that such a thing seems to be quite disgusting and gross, given the particular tale, and the relatively limited questionnaire and possible replies allowed, but in some ways it is hard to understand why it should be disgusting and maybe further experiments would clarify the situation. What I am trying to say is that these projects are often hard to understand and are not basically necessarily having a simple and clearcut follow up.

There is probably no very simple external situation either. The subjects' main interest tended to be not in (cars or whatever the gift was) but in sports and football. They did not know what the gift was, and whilst like any small child they liked presents, there was no special liking for motor vehicles, the closest connection being perhaps that one child's father was a rickshaw driver. The subjects had no known motive to 'cheat' or to make wild guesses at what they might have dreamt, or to invent dreams. Of course, mirroring the Tetris experiment, the idea was that the subjects should play a lot with the present voluntarily when they received it. They indeed did so, so at that level the technique used was cheap, modest, but apparently adequate to get a preliminary result.

Possibilities for further improvement in performance

The use of a very large number of experimental subjects of differing backgrounds and personality types could be one step forward. Another step could be the use of virtual reality apparatus during the experiment as mentioned earlier in this paper.

The use of email experiments and some form of content analysis as in Domhoff (2002, 2003) could also lead to more results though such methods become very mechanical sometimes and important detail is likely to be lost.

On that latter theme, epidemic tracking via Google (2008) is by now a commonly used method and with some variations epidemic tracking might be usable for this work. And then of course there is Google Trends, another powerful tool. Careful use of these and similar techniques could eventually be incorporated in or even replace the humble questionnaires. Less conventional internet approaches to detection as in FindTimeTravel.com have not succeeded to date.

Vul (2008) appears to take the view that cognition may be described as statistical inference, and points out that averaging reasonable guesses is better than having only one try - a result which seems to hold for one person as well as for a statistically designed goup. There are several reasons why this may be so, including the idea that the brain is continually generating hypotheses and checking them against reality. Such methods may be considered in a brain model under development. and indeed were implied in one or two earlier attempts to create a model (Yates, 2008).

Philosophical Comment

Philosophically, we are left with the interesting speculation that, using reasonable present day B-series only physics, from the work of Watanabe (1955) right up to the present day (e.g., (Gott, 1997) time travel to the past from the future as well as from the past to the future could become possible. In Gott's example, this form of time travel would be subsequent to the discovery of the first time machine. How this would apply in the case of effects relating to the reverse Stickgold effect, which we may be demonstrating in the present work, still may need to be determined. But clearly the physics might well force that Gott's condition above to apply, and we certainly cannot assume otherwise without good reason. Information theory might seem to make the restriction apply to retain consistency with current theoretical physics. But for the record, my own first patent (Yates, 1980) of time travel was made public in Patent GB2051465A during 1971 to 1979. There are plenty of potential paradoxes here for philosophers, and in my opinion, especially experimental philosophers.

114

References

Blackmore S.J., Parker J.D. (2002) Comparing the content of sleep paralysis and dream reports. Dreaming: Journal of the Association for the Study of Dreams. 12, 45-59 ; Blackmore, S. , Rose, N. , (2002) Journal of the Society for Psychical Research, 66, 29-40 ; also Blackmore, Susan (2006), http://www.susanblackmore.co.uk/ , http://skepdic.com/esp.html

Blackmore S.J.,(2005), private communication.

Baars, B.J. (2003), "How Brain Reveals Mind Neural Studies Support the Fundamental Role of Conscious Experience", Journal of Consciousness Studies, 10.

Brockmole, J. R., Wang, R. F. & Irwin, D. E. (2002) Temporal integration between visual images and visual percepts, Journal of Experimental Psychology: Human Perception and Performance 28(2):315–34.

Domhoff, G. W. (2002). Using content analysis to study dreams: applications and implications for the humanities. In K. Bulkeley (Ed.), Dreams: A Reader on the Religious, Cultural, and Psychological Dimensions of Dreaming (pp. 307-319). New York: Palgrave

Domhoff, G. W. (2003). The scientific study of dreams: Neural networks, cognitive development, and content analysis. Washington, DC: American Psychological Association.

Google , (2008) http://news.bbc.co.uk/2/hi/technoblogy/7733368.stm , http://google.com/trends http://www.theregister.co.uk/2008/11/15/google_flu_trends_privacy/print.html

Gott, J.R., Li, L. (1997), "Can the Universe Create Itself?", arXiv:astro-ph/9712344v1

Haidt, J., Koller, S., & Dias, M. (1993). Affect, Culture, and Morality, or Is It Wrong to Eat Your Dog? J. Pers Soc Psychology, 65, 613-628

Hannon, B. and M. Ruth. (1997) Modeling Dynamic Biological Systems. Springer-Verlag, New York City, New York.

Hobson, J.A. (2006) private communication with author

Hobson, J.A. (2005), "13 Dreams Freud Never Had", 83, Pi Press

Hohwy J., Frith C., (2004) "Can neuroscience explain consciousness?" Journal of Consciousness Studies, 11 (7-8): 180-198, 2004)

Jones J., Pashler H., (2007), Is the mind inherently forward looking? Comparing prediction and retrodiction, Psychonomic Bulletin & Review, 14 (2), 295-300

Malle, B.F., (2004). How the Mind Explains Behavior: Folk Explanations, Meanings, and Social Interactions. MIT Press, Cambridge, MA.

Malle, B. F., Knobe, J. (2001), The Distinction between Desire and Intention: A Folk-Conceptual Analysis. In B. F. Malle, L. J. Moses, & D. A. Baldwin (Eds.), Intentions and Intentionality: Foundations of Social Cognition. Cambridge, MA: MIT Press.

Martens, J-B. (2005), "Visual interaction", Inaugural lecture, Presented on March 18,2005 at Technische Universiteit Eindhoven, ISBN:90-386-1413-6 , Digital version: www.tue.nl/bib/

Metzinger, T. (2004), 'Being No One', MIT Press Paperback.

Milgram, S. (1974), "Obedience to Authority", Harper & Row, USA

Montague, P. R., Hyman, S. E., & Cohen, J. D. (2004). Computational roles for dopamine in behavioural control. Nature, 431, 760-767.

Nadelhoffer T., Feltz A., (2007), Folk Intuitions, Slippery Slopes, and Necessary Fictions:An Essay on Saul Smilansky's FreeWill Illusionism, Midwest Studies in Philosophy, XXXI

Pizarro D.A., Laney C., Morris E.K., Loftus E.F., "Ripple effects in memory: Judgments of moral blame can distort memory for events", "Memory and Cognition" 2006, 34 (3), 550-555

Schwitzgebel, E. (2002), Why did we think we dreamed in black and white? Studies in History and Philosophy of Science 33, 649-660.

Schwitzgebel, E. (2002a), How well do we know our own conscious experience? The case of visual imagery. Journal of Consciousness Studies, 9,35–53.

Schwitzgebel, E. (2003), Do people still report dreaming in black and white? An attempt to replicate a question from 1942. Perceptual and Motor Skills 96, 25-29.

Schwitzgebel, E. (2005), Psyche 11 (6), Difference Tone Training A demonstration adapted from Titchener's Experimental Psychology (1901-1905), vol. I, part 1, pp. 39-46

Schwitzgebel, E. (2006), Do we dream in color? Cultural variations and scepticism. Dreaming 16, 36-42.

Schwitzgebel, E. (2009), (forthcoming), Do people still report dreaming in black and white? An attempt to replicate a questionnaire from 1942. Perceptual & Motor Skills.

Segal, S.J. and Nathan, S. (1964) The Perky Effect:Incorporation of an External Stimulus into an Imaginary Experience under Placebo and Control Conditions, Perceptual Motor Skills,p.385-395

Sosa, E. (2007), "Experimental philosophy and philosophical intuition", Philosophical Studies, 132(1), 99-107.

Stickgold, R., Malia, A. & Hobson, J.A. (1999) "Sleep onset memory reprocessing and Tetris. Journal of Cognitive Neuroscience" 11(supplement)

Stickgold, R., et al , (2000), "Replaying the Game: Hypnagogic Images in Normals and Amnesics" Science 290 (5490), 350. [DOI: 10.1126/science.290.5490.350]

Stickgold, R.J., Fosse, M.J., Fosse, R., Hobson, J.A., (2003), "Dreaming and episodic memory: A functional dissociation?", J. Cogn. Neurosci. 15:1 -9.

Stickgold, R., (2005). "Sleep-dependent memory consolidation", Nature,Vol 437, p1272

Vul E., Pashler H., (2008), Measuring the Crowd Within: Probabilistic Representations Within Individuals, Psych. Science, Vol. 19, (7), 645 - 647

Watanabe, S.(1955), Reviews of Modern Physics, 27, (2), 179

Wikipedia, (2008) http://en.wikipedia.org/wiki/Link_Trainer , http://en.wikipedia.org/wiki/Flight_simulator

Yates J., (1980), Patent GB2051465A

Yates, J. (2008), http://ttjohn.blogspot.com/

Yates, J. (2008a). http://cogprints.org/6176/ , "Category theory applied to a radically new but logically essential description of time and space", PHILICA.COM, Article number 135.

Yates, J. (2008b), http://cogprints.org/6232/ , "Experimental philosophy and the MBI", PHILICA.COM, Article number 139.

4.03 Experimental philosophy and the MBI

Abstract

Various facets of the MBI are discussed, and how it can be used in connection with experimental philosophy, experimental psychology and neuroscience. Brief historical references are given. The large implications of the MBI with regards to McTaggart's paradox and the resolution of the difficulties with quantum mechanics is mentioned. Later sections deal with the mereological fallacy, multiple universes, teletransportation, mind cloning and mind splitting. Dreamwork is chosen as a prime example of the use of the MBI and recent work by Tononi and Baars is referred to.

Introduction

In this paper we deal generally with various facets of the MBI ("Many Bubble Interpretation"), (Yates, 2008), and how it can be used in connection with experimental philosophy, experimental psychology and neuroscience. I begin in section (1) with brief historical references and then proceed in section (2) to refer to the large implications of the MBI with regards to McTaggart's paradox and the resolution of the difficulties with quantum mechanics, continuing in sections (3) and (4) to deal briefly with the mereological fallacy, multiple universes, teletransportation, mind cloning and mind splitting. Dreamwork is chosen as a prime example of the use of the MBI and recent work by Tononi and Baars are referred to in that connection in section (5).

(1) History of my Contact with Experimental Philosophy and some other matters

My first contact with experimental philosophy was probably a comment in about 1967 by Ted Bastin during a meeting with Ted Bastin, and also Dorothy Emmet and R.B. Braithwaite (the 'epiphany philosophers') at a large house in Cambridge. The "Epiphany Philosophers" seemed to take it as a goal to show that christianity and science were not only compatible but that they supported one another. Further, some of their considerations of matters such as ESP could certainly be taken as pseudo-science and as such I certainly never endorsed them. It also appeared that Ted Bastin's contention at one time (Noyes , 1999) was that paranormal phenomena should be defined as contradicting physics.

From a philosophical point of view, I thought then, and still think, that there could hardly be a basic objection to making hypothetical contentions of a somewhat speculative nature as it allowed for at least a metaphorical way - if a somewhat doubtful and even sometimes probably far too naive way - to the nicing down particularities in metaphysical conjectures. Some of such conjectures could give rise to real practical concerns with the forwarding of technology, in such ways as proposed 'mind uploading' and 'mind duplication'. I go into a little more detail later in the essay, whilst considering the work of Parfit.

However. The term 'spam' was not at that time used, I believe, as the art term it seems to have become but I think Ted did not at that time altogether approve of some of the rather mechanical sounding questionnaires which are still associated to some extent with the concept of experimental philosophy and there seemed to be a feeling that the idea of experimental philosophy could well become popular, but could eventually degenerate. The effect is probably noticed as early as 1938 in the work of Naess (Naess, 1938 ; Appiah , 2007). I suppose part of the problem is that we might in effect, be throwing away the baby with the bathwater by being over zealous with some of our refinements. This must of course be avoided when it is appropriate to do so.

To sum it up, to me experimental philosophy sounded as if it might be a good idea and that it might allow such positive factors as the sharpening up, refining, and sometimes rejecting for day to day purposes, the ideas of folk philosophy.

(2) The MBI ('Many Bubble Interpretation') and its use with McTaggart's Paradox

A plan for a model of the MBI ('Many Bubble Interpretation') is described in Note 5 of Yates (2008) and elsewhere in the same paper. The MBI has already shown its utility and potential further utility as described in Yates (2008). The McTaggart paradox is regarded nowadays by many philosophers as a real paradox, which it is. Much literature is available to that effect and this will be assumed, though it can be argued in detail as has been the case elsewhere. We resolved McTaggart in Yates (2008) and the effect shown is that, to do physics or neuroscience properly, we need to bear the paradox in mind, and to use both A and B series. As an analogy, not to do so would be like pretending to live not in 3 spatial dimensions, but instead to live in Flatland. Mathematical detailing of the MBI can have a very intimate connection with the human brain and we have used a neutral monist approach, though not critically, and we tend to bear neutral monism in mind for the future. The Gestalt Bubble model of Lehar (2003) is of course not the same as the Many Bubble Interpretation (MBI), although for many years I also have been a great enthusiast of the work of Kohler and Wertheimer, as well as Lewin and Leeper and so there may be some similarities in approach.

Velmans (2003), whose model has some philosophical appeal, holds a different view from authors like Lehar (I've often referred to Velmans' work in my blog, http://ttjohn.blogspot.com/), and for this present paper I hold a similar view, up to a point. Velmans states that Lehar argues that the phenomenal world is in the brain, and concludes that the physical skull is beyond the phenomenal world. Velmans argues that the brain is in the phenomenal world and concludes that the physical skull is where it seems to be. This fits in with my own work and Velmans is also a monist. James, too, was also said to be a kind of neutral monist, as Velmans (2003) points out.

And, although Velmans (2008) very reasonably makes objections (particularly in his notes 4 and 5 and related comments) to some of the ideas put forward by Baars, the popular GWT (Global workspace Theory) model of Baars has some advantages for general use, though particular instances such as Baar's theories on magnetic fields may need to cope with some objections, and may be a little too direct. We can consider the model overall and with some refinement it can possibly cope with the thrust of my argumentation.

I will continue in Section (3) by surveying the alleged mereological fallacy and multiple universes, and then go on in Section (4) to discuss brain cloning, and in Section (5) I will briefly discuss the ongoing experiments in dreamwork, one of the many possible applications of the MBI.

(3) The mereological fallacy, multiple universes and related matters.

The mereological fallacy (Hacker, 2003) is supposedly that it isn't actually your brain that does the thinking at all. In fact, the very idea that it does is virtually incoherent: not just wrong, but meaningless. Only you as a whole entity can do anything like thinking or believing. Hacker's views are essentially based on the philosophy of Wittgenstein, which many serious thinkers disagree with entirely, or accept at best only in part as in Hohwy (2003), and in present cases relevant to neuroscience, Hohwy and Frith (2004). In fact Table 1 of Hohwy and Frith (2004) almost in itself constitutes a formula for the start of writing an experimental philosophy paper but it needs more 'aha' possibility added for an appropriate questionnaire, like the early ideas of Knobe (2003) have.

To discuss Knobe's ideas I must go into a little detail about various multiple universe conjectures.

We must distinguish plainly between firstly such ideas as the many worlds interpretations of such as Deutsch and Everett, which largely seemed to be based on the many peculiar and at times often seemingly paradoxical results which arise in relation to quantum theory; indeed the still commonly accepted Copenhagen interpretation, sometimes even described as the "shut up and calculate" interpretation, is still frequently made use of, despite its problems - indeed it was over 30 years ago that Sir Rudolf Peierls commented to me, at a meeting I had convened at the Institute of Physics, London, as to the advantages that alternative-universe approaches seemed to have over Copenhagen. Since then we have had the work of David Deutsch, Gerard 't Hooft and much other work, and hopefully this will eventually help to illumine remaining difficulties in quantum mechanics and produce other results and perhaps even allow us to specifically explicate and use in detail, at least quantum multiverses - or other and totally different approaches. But like the poor, the problems of quantum mechanics are unfortunately still with us. However up to a point we may still stand aloof from all this in our present treatment. In the MBI (or 'Many Bubble Interpretation') the Schrodinger cat problem and other such problems slot in neatly enough, and to all intents and purposes are resolved or resolvable ! And for a start quantum theory as at present described in the literature is totally B series anyway, and we know in the MBI that for a proper description of a universe we need to use an A series also.

But of course we now have a newish breed of speculated universes, namely the multiverses of such people as Tegmark, Rees and Vilenkin. Tegmark (2007) neatly classifies his universes currently in 4 levels, and roughly speaking the old style 'quantum multiverses' seem to occur in level 3. But it seems to me that, however worthy such attempts may be, they are still in my opinion very clearly within the realm of speculation, like the continuous creation theory and other theories of Fred Hoyle, and like the 'Fundamental Theory' of Eddington of previous times. Does this matter ? I think it does, as we need a serious breakthrough if any great merit is to be ascribed to such work. At the end of the day, of course there may well be some points of such theories sufficiently in touch with known reality to allow us to proceed, and of course that is important, but such a toehold in scaling the mountain of wisdom will definitely not do for all, even though we must admire the efforts of such intrepid mountaineers. Now one rider to this probable fact is that philosophical studies based on such theories can be a bit lacking. I am talking for example of Knobe's ideas on freewill and this was basically discussed by him, in Knobe (2008). The interviewer John Horgan presents the view which can lead to the idea that a relatively straightforward inflationary universe theory could be best left unused because of moral implications.

Now at this point I think it is correct to mention some brief details about the background of these two authors. Horgan it seems is an agnostic journalist increasingly disturbed by religion's influence on human affairs. His details are available on the web, and seem to be mainly in the realm of popular science. He is currently unhappy with the Templeton Foundation which he seems to feel should have been more even-handed in their funding, by awarding the Templeton Prize to someone like Richard Dawkins for example and on the other hand seems to have bet Michio Kaku that "By 2020, no one will have won a Nobel Prize for work on superstring theory, membrane theory, or some other unified theory describing all the forces of nature." Joshua Knobe is a well known philosopher whose father-in-law is Alexander Vilenkin, Director of the Institute of cosmology at Tufts University, and with whom Joshua Knobe has discussions about the universe, and indeed has published with jointly.

Basically I cannot concede to Horgan's idea that important theories (like Tegnark's or Deutsch's theories, for example) may be only 'metaphysical' in the sense that they may never have any currently acceptable proof. Also I cannot concede to Knobe's apparent idea that Vilenkin's theory, one out of many, is necessarily likely to be the right theory to follow, although Knobe (2008) himself wisely states high levels of general philosophical doubt.

It seems to me that for any theory, from my standpoint there should be a meaningful likelihood that it can be proved or disproved at some present time or within the forseeable or conceivable future. The ideas of Max Tegmark and David Deutsch, for example, look as if they are sitting there on the shelf waiting for elements of proof or disproof and whether these can be found, at some point, can be good reasons as to whether they are worthy of consideration in current physics and related disciplines. Though they could, regardless, perhaps cast meaningful shadows on the wall of philosophical speculation. Anyway the hope in present ventures is to obtain meaningful and provable results. My problem here is that some of Knobe's ideas (Knobe, 2006) on freewill would appear to be based on the philosophy of Vilenkin.

The inflationary world idea of Knobe (2006) is very clear and refers to real, observable worlds, given that Vilenkin's theory is more or less correct. It also seems to me that Knobe's theory does indeed differ from the actualism of Ayers, which is just a strong form of determinism, in some forms ruled out by way of chaos theory anyway. On the basis of inflationary theory Knobe says we may not even have "a unique copyright on our own identities" These new theoretical ideas casts up a set of new philosophical questions. Now my worry is that we are here going well into unknown territory. I for one do not accept that there is any good evidence for the inflationary world idea. Philosophers may like to speculate on it and I do not contest that idea, but a blind semi-acceptance of its truth is very much another matter. One positive possibility may be the assignment of likelihood possibilities or betting odds, so we can know how to decide how much time to give such theories or even to set up some ranking order.

For example in (Marshall, 2000) on MWI (the "Many Worlds Interpretation") it seems that "Political scientist" L David Raub reports a poll of 72 of the "leading cosmologists and other quantum field theorists" about the "Many-Worlds Interpretation" and gives the following response breakdown.

"Yes, I think MWI is true"   58%
"No, I don't accept MWI"   18%
"Maybe it's true but I'm not yet convinced"   13%
"I have no opinion one way or the other"   11%

Amongst the "Yes, I think MWI is true" crowd listed are Stephen Hawking and Nobel Laureates Murray Gell-Mann and Richard Feynman. Gell-Mann and Hawking recorded reservations with the name "many-worlds", but not with the theory's content. Nobel Laureate Steven Weinberg is also mentioned as a many-worlder, although the suggestion is not when the poll was conducted, presumably before 1988 (when Feynman died). The only serious "No, I don't accept MWI" named is apparently Penrose.

Obviously these statistics would not be easy to prepare, and probably would be much harder than simply referring to the citations index. Status of person holding the opinion and accessibility and clarity of a particular theory would be just some of the factors involved. Also it would be unwise to expect too much of new or revolutionary theories, or to base too much on impressive personalities. Joshua Knobe himself, who is seemingly becoming more and more doubtful as to which if any views on many philosophical matters are relevant or justified, seems to have begun his polls with a much less straightforward idea of how to indulge the philosophical relevance of various views, and I mention in particular the very interesting paper of Kimpe (2008) in this regard. To my mind Kimpe's (2008) paper illustrates at least one way, though it be pedagogical, as to how this sort of thing should be carried out, and of course we also have the highly inspiring early example of Knobe (Nadelhoffer, 2008). My own feeling is that a touch of Milgram (1974) could be needed for further work, and indeed in Slater (2006) the UCL group got their volunteers to wear VR helmets to experience a simulated version of the Milgram experiment. It was designed to be the same, but the strangers getting shocked were just computer animated avatars. Yet the UCL team conclude their test subjects reacted on "the subjective, behavioural and physiological levels as if it were real in spite of their knowledge that no real events were taking place." Measurements of heart rate and heart rate variability showed they reacted as though the situation was real. They were just as aware and worried they were doing wrong, but shocked the stranger anyway. Other experiments seem to have also shown much the same effect, and one could very easily be led to suppose that some feelings of grief, kinship and empathy are merely biological reactions. Without doing a study of these matters, I tend to assume that some of these reactions relate simply to the novelty of the avatar situation, and may reduce in effect the more generally experience on these matters is available. Also there is the immersive effect and the fact that the participants knew they were being watched. It is also very unclear as to the ethical situation of such experiments, and whether such methods can be used in other mass murder experiments. On the other hand these factors can be partly put aside, bearing in mind the massive and apparently largely harmless exposure of world cultures to some forms of television and video gaming.

Slater's (2007) subsequent experiments indicate that much more work needs to be done on the effect of virtual environments, but that the Slater (2006) result was no individual chance happening. Slater (2003) makes it plain that the idea of 'presence', and probably many other concepts, may need to be meticulously defined for philosophical purposes in a world which is part VR and part factual reality.

Philosophy has so far not had quite the same need of definition, and it could turn out to be very enlightening - given enough further experiments. The task is probably large but well within our control, in my view.

(4) Teletransportation and brain cloning

However now we will get down to a simple case, the one referred to by Parfit (1984) in "Reasons and Persons" as teletransportation on p199-200 of the above book. I have to say that whilst the theoretical possibilities of teletransportation may be there, using methods which have been used for atoms, the actual teletransportation of human beings or indeed any animate matter - even beings as hardy as tardigrades (Jönsson , 2008) for example - is so far not possible and for a variety of reasons may never be so.

As Parfit points out, for circumstances like the above, Wittgenstein would have pointed out "It is as if our concepts involve a scaffolding of facts .... if you imagine certain facts otherwise.....then you can no longer imagine the application of certain concepts" or Quine who advised not to "suggest words have some logical force beyond what our past needs have invested them with". But Parfit says (to paraphrase) "we strongly ... don't like" - unpleasant consequences of teletransportation. And we believe that these visceral reactions will also apply even in real circumstances. At the present state of experimental philosophy, this also seems to mean that the Wittgenstein/Quine view may be the view (i.e. we can ignore such outlandish possibilities, prior to any actual such device being in play, or as existing as a mere hypothesis) on our 'abstract' side, and the 'visceral' view may apply to our judgement for real cases (i.e. if we suppose that such a device could somehow be brought into being).

And we certainly seem to have come much closer, in recent years and since the publication of Parfit's book, to a 'visceral' situation rather than an 'abstract' one, so if we were to consider Parfit's view directly in terms of experimental philosophy, we would be perhaps wise to frame our queries in accordance with the temporal change, if we were to ask a class of aspiring young philosophers about such issues, in a rather similar way to Kimpe (2008). And indeed some of these interviews if repeated every few years might give outstandingly differing poll results.

Effect of Temporal Change on Philosophical Reasoning : But indeed we would probably for the moment at least, confine our attentions for the moment to late 20th and 21st century philosophers like Parfit, who must have been involved with the effect of temporal change (scientific change being of most importance as this can induce immediate 'no-no', but also spiritual change and political change) on philosophical reasoning over a period of time already. That of course partly accounts for the changing views of philosophers even in the 20th century, Bertrand Russell's changing (political) views on the use of nuclear bombs being only one instance out of many. And it probably also accounts in part at least for Joshua Knobe's expressed feelings for the apparent indeterminism of philosophical results, perhaps not to the level of Heidegger or Wittgenstein (Minar, 2001) who seem to try to show us how skepticism presents a symptom of our way of inhabiting our condition. They take a view as if the world had first to be stripped of the taint of meaning before it could again be rendered an hospitable environment for the dwelling of mortals.

Since Knobe takes a look at a world, constantly changing even in basic scientific and physical understanding all the time, thus leading to a continual shaking of its philosophical foundations, he is understandably not eager to take very definite views on many matters, as next week's scientific advance in (say) perception theory could undermine an entire philosophical structure of a lifetime ! Not that the proponents of such a theory will want to admit it, of course, any more than Heidegger after World War II must have really wished to truly change his philosophical lucubrations from a semi-political viewpoint to a quasi-poetical one, but circumstances, and not just physical circumstances must have made him 'want' to do so. So too Knobe has his 'abstract' and his 'visceral' and it is probably possible to take great pains to venture from there.

Time Travel in Theory and Practice

His views on such matters as freewill, morality, ethics, and even crime and punishment, seem to be up to a point conditioned by his tentative acceptance of Vilenkin's physics, and the fact that he is attempting to obtain great achievements in his thought, and indeed to relate, through experimental philosophy, his 'take' on these matters to that of others. From the present point of view a fairly all inclusive theoretical base and nonetheless a directed approach seem the most appropriate, at this time. And we bear in mind that the use of experimental philosophy may be one good way to attain this.

Perhaps obviously, deep questions of relativism arise and I try to proceed with the work by avoiding these, and also avoid a detailed consideration of the perhaps rewarding work of Gadamer on hermeneutics, or for that matter of Davidson's work.

Parfit on p203 of "Reasons and Persons" says that he considers identity as "the spatio-temporal physical continuity of an object". (One perhaps looks at Kimpe (2008) to consider the mereological aspect of this matter). One can see the relevance and explication of physical continuity, probably, but psychological continuity (and particularly continuity of memory) are also reasonably considered by Parfit at some length on p205 et seq. However in regard to something like memory, which leaves the impression of relating to the obvious physical aspects of the brain, but to an as yet unclearly defined mental feature, we seem to be right there at the coal face of understanding, and could go very wrong. We certainly do not even know accurately how to erase very specific and individually chosen memories by physical means, for example, using needles or electrodes in very small areas. Many people say that such things as thoughts and memories are spread out in some way over the whole brain, as it were like a mathematical Fourier transform. And of course Hacker goes even further than this, suggesting that it is one thing to suggest on empirical grounds correlations between a subjective, complex whole (say, the activity of deciding and some particular physical part of that capacity, say, neural firings) but there is considerable objection to concluding that the part just is the whole. Hacker then uses the traditional Wittgenstein view that to do so is nonsense. For myself I would not care to take the traditional Wittgenstein approach so far, but like Parfit would like to consider the evidence as and when it arises through neurological experiments. And Parfit's continued strivings can be looked at fairly benignly until at least p209, up to which he more or less accepts the possibility of and/or need for some sort of physical and psychological continuity.

It seems to me that in the teletransportation case of Parfit, there can be both physical and psychological continuity for both of the persons (the teletransported person and the clone), as the original person has obvious continuity and the person on Mars also has continuity in space and time, using real teleportation methods of the kind - not known when Parfit wrote his book - we have at the moment for single particles, if these can be extended to larger objects. There is a quantum effect during the creation of the clone, but this is day to day and in an emaciated way is satisfactory in B-series quantum physics, and also occurs in the MBI. The only problem is that we now have two 'originals'. A perfect chance for experimental philosophy to ask each of them questions to decide which is the authentic item, and, to me, it seems fair enough if they both seem to be authentic. But duplication is common in every production line and if people can be duplicated, both duplicants should have importance, and in fair systems, human rights.

The only 'visceral' worry to either clone should probably be as to whether he and other clones are treated fairly. Whether such a worry need occur depends on historical circumstances which are not yet with us, and which basically seem to have no immediate connection to the act of cloning

Amusingly, then, mass production could almost produce a 'detournement' effect on Heidegger's ideas. Heidegger apparently (according to Steiner) warned as early as the 1920s that 'as this soap powder [i.e. conveyor belt consumerism] spreads over the planet, over the universe, it will be almost impossible for you to be you and not just one'. On the contrary, individual 'ones' and their clones may each be able to show great individuality from a basic cloned person, and each one of us may become independently available as millions of quite separately acting individuals, totally disconnected to each other, with their own psychological reactions after the cloning. Furthermore, unlike Vilenkin's inflationary world-view where it could be said free will (given the noted restrictions) is effectively almost written out of the scheme of things in the universe, overall, there would be an almost limitless free will available for individuals to put into practice, as each clone could try something different and even compare notes. Whether there would be any means of contact between clones, other than normal physical ones available to everyone, is of course unknown. And of course such a universe could also exist in (or on top of) Vilenkin's version.

Returning now to the work of Parfit, on p273 he starts discussing mind backups. Again, this is outside of existing technology, so up to a point one could assume discussion on the matter is empty. The idea presumably is that the mind backup will simply replace the existing mind, should it become damaged, say in transit. However such a backup in practice is unlikely to have access to what happens in the mind after backup and until fatality, so is likely to to be 'just another approximate clone'. Maybe whether individuals would see such a clone to be of value could depend on their circumstances, for example family responsibilities and commitments for continuity of an existing enterprise. But it seems clear that personal continuity will have been breached.

At this point we have to bear in mind that Parfit is what he claims to be a 'reductionist' and in his understanding, persons are nothing over and above the existence of certain mental and/or physical states and their various relations. But this point in his chain of argument appears to be circular as he either claims that we can give a full description of each individual thought without assuming it has a thinker, or perhaps he is claiming that we can describe the totality of our thoughts without assuming that that has a thinker. That a person's life may be seen for some purposes as a sequence of temporal events, each one an aggregate of mental or other events, provides no grounds for the assumption that the person themselves can be identified as being this sequence. Such a sequencing may perhaps be used to tag, or keep track, of a person and even to develop theories such as the MBI, but that does not mean that we have thereby established the sequence or tagging as being the person himself. Furthermore Nagel (Harth, 2004) and others seem to define reductionism in a somewhat different way to Parfit.

(5) Use of MBI and GW theory for dreamwork

We are seeing apparent continuity of human beings' existence whilst dreams occur. Most normal people simply welcome a good night's sleep, and have no fear that it will be the end of their life, or that a new person will take their place when they 'wake up' the following day. Parfit and others have of course considered such possibilities, but we can probably rule them out for the current practical exercise.

We have already used dynamic systems modelling for the sleep condition and referred to it in Yates (2008). For the interim, rather than refining existing models we intend to try to determine empirical factors which will sharpen and enhance the reverse Stickgold effect, but to relate such progress to MBI theory also.

We note from Knobe's work that there seem to be two aspects to any real philosophical views, namely abstract ideas or views and visceral ideas or views. This suggests the possibility of something quite different to Pavlovian 'conditioning' or effects of such a nature, such as simple computer-aided classical conditioning after the manner of Richard F. Thompson. The idea is simply to give some of the subjects small rewards during the process of dream recording, likely to take place over about 10 days with the reward given mid-session. The control group will not be given rewards and we will see what difference this makes to the dreams, if any. A number of other techniques will be used during the tests, some of which are currently ongoing.

Global workspace and other theory : In connection with his theories of consciousness, Tononi (2008) claims that consciousness fades during early sleep, and that this is likely to be due to the dreamless brain either breaking down into causally independent modules, shrinking its repertoire of possible responses, or both. He has carried out work using EEG and rTMS results to help to validate this view. As Tononi puts it his theory suggests that "the brain breaks down into little islands that can't talk to one another." Now this is a reasonable postulate in the B series and whether in effect something similar occurs in the A series (which may well have additional ability to integrate information in the brain, notwithstanding B series dissociation) is as yet undetermined. Stickgold had said earlier "He has plainly and elegantly demonstrated a breakdown in the ability of cortical areas to interact normally as we fall asleep, but he hasn't provided any reason to think that this is related to the changes in consciousness as we fall asleep." "Scientists have nothing approaching an understanding of why we are conscious when we are awake or, indeed, why we are awake. So looking for what changes cause a loss of consciousness is a very difficult question because we don't know what we're looking for. I don't think this adds anything substantive about consciousness. It does add some information about the changes in brain function that accompany the shift to sleep, in a very elegant and beautiful way to show it." Tononi however seems to be of the view that "the ability of distant parts of the brain to communicate with each other constitutes consciousness."

I have by now written at length about psychological aspects of this matter - clearly a lot can be said about the hard problem , where the deceptively simple position could roughly in effect be that "the map is not the country" (or, "the Tononi effect is not consciousness"), and indeed a lot can also be said about the work of Jack et al (2007), who can also see that there are problems but would favour a different mode of tackling them. From my own standpoint, dynamic systems theory may give interesting results for the MBI, but our additional experiments could also be helpful. And, we clearly would be much happier if we could carry out mind cloning to help with the parameters (as in that case, the differences between the A series of the clones might help with parameter values, but for the moment we can certainly contrive to make do with dreams and other experimentation. But the assumption of the mere possibility of cloning could perhaps help us to better formulate the A series.

In Baars (2006) it is pointed out that Hobson and Stickgold have suggested a neural mechanism for this phenomenon in terms of cholinergic activity during REM sleep. Thus the only memory available for the recall of a dream is the small capacity WM (working memory), resulting in dream amnesia. The limited capacity of WM would yield a memory only of the final small portion of the dream. The rapid decay of WM would account for no memory of the dream at all after a slow awakening.

References

Appiah K. A., (2007), "Experimental Philosophy", Presidential Address, Eastern Division APA, December 2007

Baars B.J., Franklin S, Ramamurthy U., Ventura M. , (2006),"The Role of Consciousness in Memory" http://www.brains-minds-media.org/archive/150

Hacker P., Bennett M., (2003), "Philosophical Foundations of Neuroscience", Wiley-Blackwell, ISBN-10: 140510838X, ISBN-13: 978-1405108386

Harth E., (2004), Journal of Consciousness Studies, 11, No. 3–4, pp. 111–16

Hohwy J., (2003), Minds and Machines 13(2): 257–268, 2003

Hohwy J., Frith C., (2004) "Can neuroscience explain consciousness?" Journal of Consciousness Studies, 11 (7-8): 180-198, 2004)

Jack A., Robbins P., Roepstorff A., (2007), "The Genuine Problem of Consciousness" http://www.petemandik.com/blog/2007/01/01/pms-wips-008-anthony-jack-philip-robbins-and-andreas-roepstorff-the-genuine-problem-of-consciousness/

Jönsson K.I, Rabbow E.,Schill R.O, Harms-Ringdahl M., Rettberg P., (2008). "Tardigrades survive exposure to space in low Earth orbit",Current Biology, 18

(17): R729-R731. doi:10.1016/j.cub.2008.06.048.

Kimpe K., (2008) ,
http://experimentalphilosophy.typepad.com/experimental_philosophy/2008/09/
polling-as-peda.html

Knobe, J. (2003). "Intentional Action and Side Effects in Ordinary Language."
Analysis, 63, 190-193 ; and   http://experimentalphilosophy.typepad.com/
experimental_philosophy/files/knobe_writeup.doc

Knobe J., Olum K., Vilenkin A., (2006), "Philosophical Implications of
Inflationary Cosmology", The British Journal for the Philosophy of Science,
57(1):47-67; doi:10.1093/bjps/axi155

Knobe J., (2008), http://bloggingheads.tv/diavlogs/8796

Lehar S., (2003) "Gestalt Isomorphism and the Primacy of the Subjective
Conscious Experience: A Gestalt Bubble Model. Behavioral & Brain
Sciences" 26(4), 375-444

Marshall J., (2000),
http://www.themilkyway.com/quantum/FinalReport/IntroductionQE.html

Milgram S., (1974), "Obedience to Authority", Harper & Row, USA, ISBN: 0
422 74580 4

Minar E, (2001), "Heidegger, Wittgenstein, and Skepticism", Harvard Review
of Philosophy, Vol IX, p37

Nadelhoffer T., (2008), http://experimentalphilosophy.typepad.com/
experimental_philosophy/2008/06/the-knobe-effec.html

Naess A., (1938), "'Truth" As Conceived By Those Who Are Not Professional
Philosophers' (Oslo: I Kommisjon Hos Jacob Dybward, 1938)

Noyes, H.P. (1999) arXiv:quant-ph/9906014v1 3 Jun 1999

Parfit D., (1984), "Reasons and Persons" Oxford Clarendon Press, ISBN 0-19-
824908-X

Slater M., (2003) , " A note on presence technology"

http://presence.cs.ucl.ac.uk/presenceconnect/articles/Jan2003/melslaterJan272
00391557/melslaterJan27200391557.html

Slater M., et al. (2006) "A Virtual Reprise of the Stanley Milgram Obedience
Experiments", PLoS ONE 1(1): e39. doi:10.1371/journal.pone.0000039

Slater M., et al (2007),The International Journal of Virtual Reality, 2007,
6(2):1-10 ; and Geser H., (2007), "A very real Virtual Society. Some
macrosociological reflections on "Second Life"". In: Sociology in Switzerland:
Towards Cybersociety and Vireal Social Relations. Online Publikationen.
Zuerich, May 2007 http://socio.ch/intcom/t_hgeser18.htm ; and Hagni K. et al
(2008), "Observing Virtual Arms that You Imagine Are Yours Increases the
Galvanic Skin Response to an Unexpected Threat". PLoS ONE 3(8): e3082.
oi:10.1371/journal.pone.0003082

Tegmark M., (2007) arXiv:0704.0646v2 [gr-qc]

Tononi G, Massimini M. , (2008), "Why does consciousness fade in early
sleep?", Ann N Y Acad Sci 2008;1129:330-4.

Velmans M., (2003) "Is the world in the brain, or the brain in the world?" (A
commentary on Lehar, S. "Gestalt isomorphism and the primacy of subjective
conscious experience: A Gestalt Bubble model", Behavioral and Brain
Sciences, in press). [Journal (Paginated)] (Unpublished) ; cogprints.org/2756/

Velmans M., (2008), "Reflexive Monism",
http://www.goldsmiths.ac.uk/departments/psychology/staff/velmans.html ,
Journal of Consciousness Studies (2008 – in press)

Yates J., (2008), Philica no 135 ; http://cogprints.org/6176/

Time Travel in Theory and Practice

5

The MBI and its involvement with Quantum Theory and the rest

Chapter 5 begins with a description of the way the MBI describes how the quantum theory relates to representations of the A series. It then proceeds to resolve the Schrodinger cat problem using the MBI. The MBI does not of course propose to be a replacement for quantum theory but its existence may help to clarify the problems brought up in quantum theory, just as the many worlds interpretation is not normally said to be a replacement for quantum theory. In Sections 5.03 and 5.04 we proceed to describing various experiments which can be, or have been, used to adjust the mind in accord with the MBI.

5.01 Episodic Memory, Cognitive Dissonance, and the MBI Bubble

Egan (2007) has recently carried out a series of experiments which could be taken to suggest that humans and monkeys have similar cognitive dissonance responses. i.e. children and monkeys apparently tend to change current preferences to fit past decisions. In Egan's experiments, for example, the subjects had first assessed two possibilities as of equal value, and then had to make a decision based on the idea that they were not. Then they derogated unchosen alternatives. This seems to lead to the idea that the core mechanism of cognitive dissonance is simple as very young children and monkeys do it.

Enthusiasts of the principle of cognitive dissonance (Stafford, 2007) often assume complex mechanisms.

But we do not have any decent evidence that the change in attitudes or

135

'dissonance' comes from an uncomfortable mental state or that it arises from a contradiction between beliefs and actions. Indeed, it would seem that not only do you need to have abstract beliefs about the world and yourself, you need to have some mechanism which detects when these beliefs are in contradiction with each other or with your actions, and which can (unconsciously) adjust selective beliefs to reduce this contradiction.

Secondly, all this sophisticated mental machinery is postulated to exist from changes in behaviour, but it is never directly measured. But for young children and monkeys, that all seems rather complicated. Surely a much simpler process of choice is somewhat like the idea of the mind operating like a basic computer in naive cases. A decision between A vs B is not just "choosing A" but is also "not choosing B".Then, when the choice becomes B vs C, even a naive mind is more likely to choose C because it is simply re-applying the previous decision of "not choosing B", rather than performing some complicated re-evaluation of its previously held attitudes a-la cognitive dissonance theory. These choices include some statistical white noise, so ultimate results for a choice about C are going to reflect that, of course. Children and monkeys both grow and evolve, and one way this might happen is that the increasing complexity of neural pathways during this process could add perceived elements to the mind which might be construed by it as chagrin or whatever appropriate feeling is relevant. This too, may be relevant in some ways to discriminatory process but from the work of Egan does not seem necessarily to be at the core of the adult, sophisticated discriminatory process. The advantage of the latter is that it might lead to better and more sophisticated decisions for reasons such as personal or species survival.

Interpretation is not neutral, but represents a statement to the world as to the stance it implies. For example, chagrin at a perceived wrong first opinion (which in the simple example above would be a preliminary perceived idea that A was substantially the same as B), may alter to puzzlement if the same situation arises again and again - say in respect of some D,E and F. An example is in Eisenstein (2007) where a lecturer plausibly debunks a largish group of students about their several views as expressed in individual essays. Eventually they begin to protest at his alleged scepticism. This probably at least illustrates the high level of sophistication they have reached, and on the face of it suggests that more and better window dressing is needed for any theory of mind than we get directly from cognitive dissonance.

This really seems to amount to a similar level of discrimination to that of the recognition of sarcasm. Now we have some neurological evidence as to how such levels of discrimination operate. Shamay-Tsoory (2007) found that people with prefrontal damage had trouble recognising sarcasm, but people with damage in posterior brain areas were unaffected. People with damage in the right hemisphere and the prefrontal lobe (most profoundly evident in those with right ventromedial lesions) also had problems understanding the emotional cues involved in processing sarcasm, such as tone of voice, which correlated with their ability to understand sarcasm.

It was construed that the brain's language areas interpret the literal meaning of a statement, the right hemisphere and frontal lobes process the emotional context, while the prefrontal cortex integrates the two.

The total sample size was 25. This suggested to the experimenters that the right frontal lobe mediates understanding of sarcasm by integrating affective processing with perspective taking.

Now consider episodic memory. Recent work (Rosenbaum, Tulving, 2007) seems to show that theory of mind (ToM) can operate independently of episodic memory. ToM is doubtful in autistic behaviour (Baron-Cohen, 1985) and some doubt it even for non-mammals (Ramsay, 2007). Rosenbaum also says that ability to detect sarcasm , deception, and similar attributes, are not removed by such memory loss.

Further, total loss of personal memory made no difference in subjects tested. Indeed Rosenbaum goes so far as to say: "We found that if you're trying to put yourself mentally in someone else's shoes, you don't need to put yourself in your own shoes first."

On that basis, which comes from practical experiments actually carried out with only some fairly basic and probably defensible propositions, we do not need of necessity a very complex ToM to cover a great deal of human behaviour. Even episodic memory of episodes relevant to cognitive dissonance, can perhaps be left out of a preliminary model which deals with cognitive dissonance !

We can even work out which parts of the brain may be relevant to such a proceeding. For episodic memory the relevant regions may be regions within the medial temporal lobe (MTL), such as the perirhinal cortex, parahippocampal cortex and entorhinal cortex (Moscovitch et. al. 2006).
Now the above results on cognitive dissonance and episodic memory seem to indicate that we can set up a pretty good Stella or Madonna model for human

behaviour without taking account, at least in a preliminary (highly foundational) way, of a complex ToM (Ramsay, 2007).

So a relatively simple basic mathematical model for a "bubble" in the MBI ('Many Bubble Interpretation') discussed earlier, can be constructed. Each bubble will be much the same, in principle, as given in the mathematical Madonna descriptions given in earlier blog entries herein. And in the simpler cases there need not exist episodic memories to retain many of the apparently intrinsic features of human thought. We can even, in terms of level of simulation simplification, try to emulate Winfree (e.g. in Izhikevich, 2007).

Now there is no need to deal at this juncture with the problems posed by Honderich (1984) or by, for example Trevena (2002) and others, to the work of Libet (2003) and its defence by Haggard (2005), Klein (2007) and others. Libet's results, or others, will just be part of the Madonna formalism within the bubble, which can be "pseudo A series" in its formulation, I think. Further experimental work will follow.

References

Baron-Cohen, S., Leslie, A.M., & Frith, U. (1985) Does the autistic child have a 'theory of mind'? Cognition, 21, 37-46

Egan L. C., Santos L.R., Bloom P. (2007), "The Origins of Cognitive Dissonance: Evidence from Children and Monkeys". Psychological Science, 18, 978-983.

Eisenstein C., (2007), http://www.realitysandwich.com/node/845

Haggard, P. (2005). Conscious intention and motor cognition. Trends in Cognitive Sciences 9, 6 : 290–295.

Honderich T., (1984). "The time of a conscious sensory experience and mind-brain theories", J. Theoretical Biology 220:115-119. ; "Is the mind ahead of the brain? -- Benjamin Libet's evidence examined"; "How Free Are You? The Determinism Problem" and many others.

Izhikevich E.M.,(2007)"Dynamical Systems in Neuroscience",ISBN 978-0-262-09043-8, M.I.T.Press reprint, http://vesicle.nsi.edu/users/izhikevich/publications/dsn/index.htm

Klein S.A., (2007) "Do Apparent Temporal Anomalies Require Nonclassical Explanation?",
http://cognet.mit.edu/posters/TUCSON3/Klein.html ; "Libet's Timing of Mental Events", Consciousness and Cognition 11, 326–333 (2002) doi:10.1006/ccog.2002.0569

Libet B.,(2003). "Can Conscious Experience affect brain Activity?", Journal of Consciousness Studies 10, nr. 12, pp 24-28 ; and many others.

Moscovitch M., Nadel L., Winocur G., Gilboa A., Rosenbaum R.S. (2006) "The cognitive neuroscience of remote episodic, semantic and spatial memory" Curr Opin Neurobiol. 16(2):179-90.

Ramsay T.Z. , (2007), Science and Consciousness Review, November 26, 2007

Rosenbaum R.S.,Stuss D.T.,Levine B., Tulving E.,(2007),"Theory of Mind Is Independent of Episodic Memory", Science, 23 November 2007: Vol.318.no.5854, p. 1257 DOI: 10.1126/science.1148763

Shamay-Tsoory S.G.,Tomer R., Aharon-Peretz J., (2007), Neuropsychology (vol 19, p 288)

Stafford T., (2007),
http://www.mindhacks.com/blog/2007/12/cognitive_dissonance.html

Trevena J.A., (2002),Conscious Cogn.,Jun;11(2):162-90; discussion 314-25.

5.02 Many Bubble Interpretation (M.B.I.)

The MBI ('Many Bubble Interpretation')

The 'Many Bubble Interpretation" appears by means of a model of the McTaggart A series. Without being initially sidetracked into the fascinating coherentist theories of epistemic justification, we simply loosely define A series bubbles for present purposes as being entities inside which a person, persons or whatever are for the moment severally confined, each at some personal present (which we know from as far back as the work of Kornhuber, Libet, etc., is not readily defined as a single point in time, but more usually is taken by psychologists and others to have at least some ongoing 'duration'), and with a past, a present and a future, in accord with the spirit of the McTaggart A series. The work of LePoidevin, Quentin Smith, Dean Zimmerman and many others is borne in mind. And as Dyke has said, we may not be forced to countenance plurality of further worlds in such circumstances – although we can. The A series is treated as a large category, intrinsically unmappable one to one onto the B series. There is also a B series and this can often be represented by a quantum mechanical description of the universe.

So we have both an A series and a B series, and McTaggart's work and Zeno's work, (and/or their modern counterparts), can pose no problems.

Outline of the MBI

A relatively simple basic mathematical model for a "bubble" in the MBI ('Many Bubble Interpretation') discussed earlier, can be constructed. Many bubbles – and there would be many – could be much the same, in principle, and given by Berkeley Madonna, for example. And in the simpler cases of the model there need not exist episodic memories to retain many of the apparently intrinsic features of human thought (Egan, 2007). Even total loss of personal memory made no difference in subjects tested. Indeed Rosenbaum (2007) goes so far as to say: "We found that if you're trying to put yourself mentally in someone else's shoes, you don't need to put yourself in your own shoes first." We do not even need, of necessity, to consider mirror neurons to 'have a life'. We can even, in terms of level of simulation simplification, try to emulate Winfree. And no complex 'Theory of Mind' is required (Ramsey, 2007).

There is no need to deal at this juncture with the problems posed by Honderich or by, for example Trevena and others, to the work of Libet (2003), and its defence by Haggard, Klein and others. Libet's results, or others, will just be part of the Madonna formalism within the bubble, which can be "pseudo A series" in its formulation, I think.

Obviously, more complex contents can be given to the MBI and this is being done.

References
Egan L. C., Santos L.R., Bloom P. (2007), "The Origins of Cognitive Dissonance: Evidence from Children and Monkeys". Psychological Science, 18, 978-983.

Libet B.,(2003). "Can Conscious Experience affect brain Activity?", Journal of Consciousness Studies 10, nr. 12, pp 24-28 ; and many others.

Time Travel in Theory and Practice

Ramsay T.Z. , (2007), Science and Consciousness Review, November 26, 2007

Rosenbaum R.S.,Stuss D.T.,Levine B., Tulving E.,(2007),"Theory of Mind Is Independent of Episodic Memory", Science, 23 November 2007: Vol.318. no.5854, p. 1257 DOI: 10.1126/science.1148763

Applications of the MBI (examples only)

In Quantum Theory:
Quantum Computing – even for Paul Kwiat's work – easier to understand in real terms (a)
The 'Schrodinger cat paradox' - not such a paradox any more (b)
Schrodinger's kittens – also fall in line (not literally !) (c)

A Specific Application:
A problem involving some applied mathematics and philosophy. (d)

The Many Bubble Effect described herein, together with other factors like McTaggart's paradox and Zeno's paradox, allowed a formulation in terms of differential equations of Stickgold's dream experiments and my interpretation and furthering of them. This led to a number of equations and graphical results. In particular to equations like that described as N003b on my website at http://ttjohn.blogspot.com/ (RSS feed available) and on the CD, obtainable here.

Very briefly, as the 'pseudo A series' might describe it, there could be tiny pushes and impulses to the mind at a given time, from both past and **future** stimulations, but at a particular time it could be said that the mind is in some kind of dynamic balance which Stickgold altered in the 'Tetris dream' by a push from the past, relatively easy in retrospect. In my case I alter the position of the push from the future to the present, and this worked too. Experiments and trials are still under way, and could show conclusively the merits of the MBI, though their success is not essential to it.

Important Next Step – Could perhaps be done by anyone !

The next step may well involve the refinement or replacement of the present equations in Berkeley Madonna using methods of Self Organised Criticality, and in particular the use or incorporation of a model like the sandpile model may help.

The simplified form of the equations in the present model, described in more detail on the website, was

$dR/dt = a*R + b*J*(1-|J|) + e*Z$
$dJ/dt = c*R*(1-|R|) + d*J + f*Z$
$dZ/dt = h*S + g*R$ (S is Heaviside step functions : in N003a, g=f=0)

If these equations can be improved and/or accurate limits set on their parameters, they could be used for yet more tests and even more accurate results, in for example the mode, duration and timing of stimuli. (We might well bear in mind Winfree's work as a parallel example of such methods). Very roughly, R ('Romeo') and J ('Juliet') represent the 'unconscious' and 'conscious' mind or equivalent representations in other philosophical approaches, and Z the applied impulse.

{The paper "Self-Organised Criticality – a possible tool for the MBI" shows some of the present basis of that idea, the paper "Episodic Memory, Cognitive Dissonance, and the MBI Bubble" fills in some more of the details. "McTaggart's A and B series and how they relate to modern science and neuroscience in the 21st Century" briefly describes the relevance of McTaggart's work. "2007 TSC & Qmind Conferences: My Abstracts & brief Details" give earlier posters at Salzburg and Budapest conferences and also provide brief abstracts of supporting work including equations and graphs, all on the CD. A further abstract has been accepted at the Arizona conference later this year. "Human Consciousness, Philosophy and Computation" is a fairly comprehensive summary of the position at that point. All on website http://ttjohn.blogspot.com/ }

Time Travel in Theory and Practice

Schrodinger's Cat

Copenhagen etc. - leaves usual unexplained problems   M.B.I.   explains and progresses further work

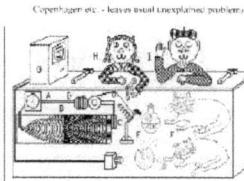
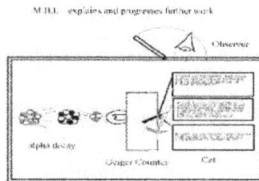

Simple Tetris game illustrates M.B.I. application – others to be considered

A   *In the experiment on the left, the subject plays Tetris and then dreams about it. On the right, the time positions are reversed.*

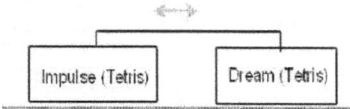
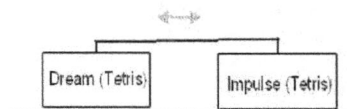

"Time" ►   "Time" ►

Stickgold Effect   **Reverse Stickgold Effect**

B   *Dream*   *Impulse*

Dr. John Yates,
email: uvscience@gmail.com
Institute for Fundamental Studies, Vasai, Mumbai, India 2007-2008.

M.B.I. – explains and progresses further work

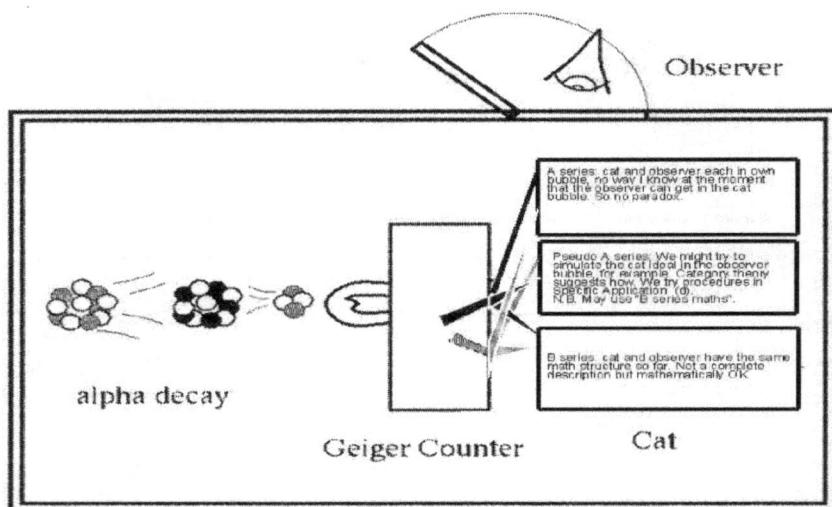

alpha decay

Geiger Counter

Cat

Observer

A series: cat and observer each in own bubble, no way I know at the moment that the observer can get in the cat bubble. So no paradox.

Pseudo A series: We might try to simulate the cat ideal in the observer bubble, for example. Category theory suggests how. We try procedures in Specific Application (d).
N.B. May use "B series maths".

B series: cat and observer have the same math structure so far. Not a complete description but mathematically OK

Copenhagen etc. - leaves usual unexplained problems

5.03 Quantum Interrogation, the McTaggart A Series, and the Many Bubble Interpretation

Abstract

The 'Many Bubble Interpretation" appears in a model of the McTaggart A series. Without being initially sidetracked into the fascinating coherentist theories of epistemic justification, we simply loosely define A series bubbles for present purposes as being entities inside which a person, persons or whatever are for the moment severally confined, each at some personal present (which we know from as far back as the work of Kornhuber, Libet, etc., is not readily defined as a single point in time, but more usually is taken by psychologists and others to have at least some ongoing 'duration'}, and with a past, a present and a future, in accord with the spirit of the McTaggart A series. The work of LePoidevin, Quentin Smith, Dean Zimmerman and many others is borne in mind. And as Dyke has said, we may not be forced to countenance plurality of further worlds in such circumstances - although we can. The A series is treated as a large category, intrinsically unmappable one to one onto the B series. There is also a B series and this can often be represented by a quantum mechanical description of the universe.

I start with a brief explanation of the idea of quantum interrogation as clearly the relevance of quantum theory to the mind has great relevance. This fact was noted at a very early date in the so-called 'Schrodinger Cat Paradox'. I attempt to retain the 'Cat paradox' here, in my new Many Bubble Approach, but in a way that is helpful and warning in a kindly way, rather than minatory and implying the possibility of immediate muddle and paradox - a use for which the 'Cat paradox' seems to have been frequently historically put. It transpires that when used with the MBI, that according to Kwiat's interpretation of his work on quantum optics, for the purposes of computing by a quantum computer, it should be possible to almost noninvasively study the human mind, probably in a way at least as noninvasive as fMRI scans. In explaining this, the illustration given by Dean Carroll about measuring the presence of a sleeping puppy without waking him up is considered, as well as other aspects of the quantum interrogation matter.

Further, there are other useful applications of the MBI, in particular for dream research and perhaps many varied psychological experiments such as near death experiences and synaesthesia. Work is proceeding at the Institute for Fundamental Studies, Vasai, near Mumbai, India. The dream research experiments are not construed as precognition but as an application of an advanced Stickgold effect.

Introduction

There is a brief discussion of quantum interrogation followed by its application to the effect on the McTaggart A series, in a way which also expands the description of the A series given on 4th July 2007 and on 22nd September, 2007, at http://ttjohn.blogspot.com/. The Many Bubble Interpretation (MBI) is introduced. A rough notional illustration shows how in principle we can detect the presence of a sleeping puppy in a box, using quantum theory, without disturbing the puppy by more than an almost infinitesimal amount. The MBI also allows a better understanding of the Schrodinger Cat paradox, and so on. An obvious potential application to neuroscience, using the A series, is referred to in the conclusion of the article.

Brief explanation of Quantum Interrogation

Quantum interrogation appears a stranger phenomenon than most which are found using quantum mechanics. Press reports have frequently suggested that it is even stranger than it is.

Kwiat says roughly the following: Sometimes called interaction-free measurement, quantum interrogation is a technique that makes use of wave-particle duality (in this case, of photons) to search a region of space without actually entering that region of space. Utilizing two coupled optical interferometers, nested within a third, Kwiat's team succeeded in 'counterfactually' searching a four-element database using Grover's quantum search algorithm.

Briefly, the Press in effect have sometimes omitted the word "entire" from a paraphrased version of the statement apparently made by Professor Kwiat "It seems absolutely bizarre that counterfactual computation – using information that is counter to what must have actually happened – could find an answer without running the entire quantum computer." The word "entire" is there to stay in my view, but needs explaining - and the explanation involves the almost equally bizarre Quantum Zeno effect.

A lack of understanding of what happens has led to the following explanation by some (not I). Say you've got two programs running, P1 and P2. P1 is performing some enormous calculation, while P2 is doing nothing. If P1's calculation returns any answer other than 5, then P1 closes P2. You come back to your computer and find that P2 is still running. Even though P2 didn't calculate anything, and even though P1 never did anything to P2, you can immediately conclude — from P2 alone — that the answer you wanted was 5.

But that is not how it happens, and it is important to know that.

The simple "sleeping puppy" explanation in Carroll (2006) seems to me to roughly explain the matter.

Time Travel in Theory and Practice

A puppy is allegedly sleeping in a box. We wish to make sure he is in the box without opening it. If we pass dog food to him through a slot in the box, he will wake up noisily and eat it, and we will know he is still in the box. If we pass in a salad, the puppy will stay asleep.

We do not want to wake the puppy. To avoid this, we use quantum theory. Let us suppose we can describe the puppy's food quantum- mechanically in a simplistic way, as in the equation of Carroll (2006)

food = a* salad + b*dogfood The components of the vector are a and b. .... (1)

In *Note 1* we briefly explain how in these circumstances, we can detect the presence or otherwise of the puppy in the box with an extremely small (almost as small as we want) chance of waking him. It seems to me that, in *Note 1*, the first time the wave function is rotated by a small angle it will collapse if observed by the puppy, under the Copenhagen interpretation. The main problem in Note 1 is that Note 1 requires a good deal of removal and insertion of the food into the box, having the experiment set up in such a way that it involves quantum mechanics, and so on, or in real life terms, for example something like carrying out the Kwiat (2006) experiment which involves fairly normal experimental equipment in quantum optics.

Interpretations of the Kwiat experiment still seem to be at issue, Kwiat (2006a) and Mitchison (2006). Furthermore, it seems to me that to understand the fully the application of the quantum Zeno effect in the Kwiat approach, one has to consider the actual Kwiat experiment as described, and indeed the hypothetical Kwiat experiment and not go to the simple illustrative example above and in Note 1. In Carroll (2006) there is reference to Kwiat's own blogged comments, which should help somewhat. Kwiat (2006b) gives Kwiat's own account of the Quantum Zeno effect.

Relation to the A series

I have written several times about the McTaggart A series so will only briefly recapitulate. There are numerous interpretations of the A series, presentist (rather like actualist in metaphysics), 'Growing Block' (almost explanatory), eternalist (for example perhaps Smith, Craig, and Williamson) and in my view a few other types. Zimmerman (2005) mentions some variants of the A series.

Newton-Smith mentions "We [may] lack the grounds to assume time has its topological properties as a matter of necessity ... It is [perhaps] the task of the philosopher to demonstrate the consequences of supposing it to have such and such a topology". Quite so. For immediate purposes we will not even consider where, if at all, Swinburne and others fit in with all this. And the work of Dyke (1998) makes it clear that we may be dealing with models only in such circumstances and may not be forced to countenance further plurality of worlds.

Indeed, there seem to be more 'metaphysical' variations of the A series, in existence or possible, than there are of a 1958 Ford Edsel, but hopefully not of the same fate.

So to keep it simple for the moment, I am provisionally simply copying from Maxwell (2004) the rough definition, which we may change if fitting "McTaggart, famously, distinguished two conceptions of time: the A-series, according to which events are either past, present or future; and the B-series, according to which events are merely earlier or later than other events".

Now Baldwin (2004) says "This point connects with a deep distinction between practical and theoretical points of view. The practical point of view is essentially 'first-person' ('subjective'): it assumes knowledge of who I am (TRB), where I am (York), what time it is (today's date). The theoretical point of view is essentially impersonal ('objective'). It doesn't require this first-person knowledge. So the A-series/B-series distinction is a case of the distinction between these two points of view, the practical and the theoretical."

Or to put it my way, we would probably be talking about a person in the A series, a map in the B series, and whilst we leave aside Baldwin's idea that the A series is practical but the B series theoretical, which is pretty much an exact fit so far, it does follow that my brief definitions in http://ttjohn.blogspot.com/ that the A series is good for people but the B series is best for objects, is actually more mainstream than some of the bizarre metaphysics in the literature, though as with all mildly homespun (some prefer the term foundational) ideas it does need watching closely.

Of course we need not bother with all this, and just stick with the B series or something like it that will satisfy our problems in a half-hearted way, but then we seem to be left with a 'theoretical' result that looks sort of funny, with a Schrodinger's cat and kittens and many other things unexplained, with no clear chance of getting at humanity or even a God or Gods if we believe them to be there, no direct link to biophysics or psychology or even the study of dreams and other as yet rather strange but undoubtedly existing phenomena, and perhaps worst of all, what distinctly looks like an unresolved case of McTaggart's paradox, implying perhaps the non-existence or vacuous nature of time itself, unresolved metaphysical puzzles and all the rest of it. Much more can be said but why strain ? My way can lead to a simple and constructive expansion of the philosophy of the mind. And of course time does involve change, we know from personal experience that it does and that people have a past, a present and a future, and allowing change is the essence of the MBI approach.

In my mind, and not entirely accurately, I tend to think of the A series as being like a lot of bubbles floating freely, each of which representing a person or sentient object, and his or her past, present and future at some time, and we could hopefully index the persons in the bubbles as (Pn,Tm), this being person Pn at time Tm. By now the apparition has degenerated to a pseudo A series (almost nearer to being a B series). But in principle we are mapping an A series to some model we can understand. And if we want we can follow memes through the bubbles by now, and index like (Pn,Tm,Mi) where Mi is some meme which may occur as part of one or more bubble. But this is intended as a guide rather than mathematics or metaphysics.

And whilst as presented above, the A series has a "future" along with each "present" and "past", in the individual bubbles. This is only a model and not a metaphysical description of the universe. It is however by the nature of the model, many world in structure. The claim is not made that these many worlds have to exist in actual fact. So the MBI ("Many Bubble Interpretation") seems to be in basic distinction from the MWI of Price or Deutsch or indeed the MCI ("Many Computations Interpretation") of Mallah (2007). The latter two are in origin B series, and to aid consistency should possibly be assumed to exist, in some sense, at least in the sense that the quantum mechanical results in Hilbert spaces exist. In the MBI, the many bubbles of time, each with its own past, present and future, are as real as the person or conglomerate observing them, and only exist in a model of the A series. The A series itself, in some metaphysical sense at least, can be taken to exist. So the Baldwin (2004) bubble referred to above, will contain the person (TRB) at the indexed time in the bubble in York, with a past somewhere else, perhaps partly in Leeds, and presumably a future somewhere else again, perhaps partly in Blackpool. This will simply be at the 'time' referred to above for TRB, and the bubble is only part of a model which contains many more bubbles. But this is only part of a model of the A series.

And, without even invoking quantum theory, Quentin Smith (2002) explains how some models of the A series can seem to have B series concomitants, even in special relativity. In fact if we wish, we could consider our A series bubbles corresponding to different Tm values to be linked to one another by a spider web of gossamer chains. The spider web could now seem to be very clearly savouring of the B series, although we had started with a model based on the A series. Given suitable provisos, that spider web might well suit STR (Special Theory of Relativity).

Furthermore, the Schrodinger Cat riddle seems to give no essential problems in the MBI, and the MBI has the additional virtue of flagging up the obvious apparent anomalies that the Cat paradox has seemed to show to some, in the B series (*Note 2* gives more details).

At any rate, the important factors which may distinguish the A series from the B series, both being discussed in some detail in http://ttjohn.blogspot.com/ , are likely to be in the simple supposition of use referred in this note. That the situations involving human interaction are likely to be best described in the A series. Since the human/puppy experiment mentioned above will entail human and puppy observations, and/or possibly contrived mathematical situations simulating or paralleling these, we will see in the A series, the 'counterfactual' (?) results of these experiments - in a reasonably good model of the circumstances at the corresponding A series point. In other words, much what Copenhagen theorists would predict from a quantum-mechanical Kwiat experiment. The math may occur in the B series, but the human (and possibly the puppy) observers would emerge in a good A series model. So the strange thing of Kwiat will be intrinsic to the B series but only be necessary in terms of results in the A series model - which could still probably drill down deeper by a mathematical formalism which would parrot the B series. And of course we must remember that we are really dealing with a pseudo A series in formalisms devised for the B series. The interesting thing is that here the A series will represent the stolid physical truth, in a provisional model, but the B series will look mildly paradoxical in human terms. But the system has now been tamed to the point where the apparent absurdities in quantum mechanics now sound no more unusual than the use of, say, complex numbers in algebra. And importantly there is now a link between quantum mechanics and consciousness.

And, we can use different A models for different circumstances Swinburne's evangelising can, at a pinch, fall under our aegis. And models being what they are, there is nothing unusual in using new models for new positions within the A series, if we need to change our description of the A series somewhat as time goes on.

Conclusion

The fairly simple discussion above shows that with the A/B series differentiation I have discussed elsewhere we can avoid worrying too much about strange quantum results, and concentrate on obtaining new biophysics results as I have done in the http://ttjohn.blogspot.com/.entries on 4th July 2007 22nd September, 2007 entries. In the long term we will thus probably be able interpret latest quantum results in a totally non-paradoxical way, which is even true, or may be true if Kwiat is correct, for counterfactual quantum interrogation. MBI is clearly not simply Copenhagen, MCI or MWI, but a description which can create models and building blocks capable of being better than all of these and can incorporate all or any of them, and humans as well.

Further applications of Quantum Interrogation: In the simple example above, we could replace the sleeping puppy by a human brain, in principle, and thus be led to a wonderland of strange quantum mental applications, with which, as with fMRI scans, we might hope not to damage or alter the brain. Now generally speaking, whether we all end up agreeing with Kwiat or not, the puppy seems to have been interrogated while still asleep, and this makes a point anyway. If sticklers query that as well, I need only refer to Smith's (2002) comments on well-established special relativity theory and point out that our MBI interpretation will smooth over that problem in the A series enough to show MBI's potential utilisation. So whilst it will be much happier if Kwiat is right, in no way is that matter intrinsic to the continued merit of the MBI.

Acknowledgments

My thanks to Professor Kwiat for providing a copy of "Counterfactual quantum computation via quantum interrogation".

References

Bouwmeester D., Ekert, A., Zeilinger A., (2000), "The Physics of Quantum Information", pp146, 159, Springer.

Carroll S., (2006), "Quantum Interrogation"
http://cosmicvariance.com/2006/02/27/quantum-interrogation/ ; Kwiat's own
comment is at http://cosmicvariance.com/?p=674 and fully covers his position
up to the point of writing, as far as I know.

Dyke H., (1998), pp93-117, "Real Times and Possible Worlds", in "Questions
of Time and Tense", LePoidevin , Oxford University Press, Oxford, England
OX2 6DP.

Kwiat P.G. et al, (2006), Nature 439, 949-952 (23 February 2006)
doi:10.1038/nature04523 ; there is a very brief description at many other
places, for example http://www.primidi.com/2006/02/23.html

Kwiat P.G. et al, (2006a), "Counterfactual computation revisited", arXiv:
quant- ph/0607101 ; "Weak measurements and Counterfactual computation",
arXiv:quant-ph/0612159v1

Kwiat P.G. et al, (2006b), "The TAO of Quantum Interrogation",
http://www.physics.uiuc.edu/people/Kwiat/Interaction-Free-
Measurements.htm

Mitchison G., Jozsa R. , (2006), "The limits of counterfactual computation",
arXiv: quant- ph/ 0606092 ; "Sequential weak Measurement" , arXiv:
0706.1508v2 [quant- ph]

Mallah J.,(2007), arXiv:0709.0544 "The Many Computations Interpretation
(MCI) of Quantum Mechanics".

Maxwell N., (2004), "The Ontology of Spacetime", Conference in Montreal,
14 May 2004

Newton-Smith W.H., (1980), "The Structure of Time", 61-66 inter alia,
Routledge and Kegan-Paul.

Penrose R., (2000), "The Mysteries of Quantum Physics" in "The Large, The
Small, and the Human Mind", p70 onwards, Canto, Cambridge University
Press.

Penrose R., (2005), "The Road to Reality", p804 onwards, Vintage Books,
London.

Smith Q., (2002), "The incompatibility of STR and the tensed theory of Time", Published In: "The Importance of Time", editor, L. Nathan Oaklander. Kluwer: Philosophical Studies Series.

Zimmerman D.W., (2005), "The A-Theory of Time, The B-Theory of Time, and 'Taking Tense Seriously'", Dialectica Vol. 59, N° 4, pp. 401–457

Note 1

This is largely due to Carroll (2006). Start with some food in the (salad) state. Stick it into the box; whether there is a puppy inside or not, no barking ensues, as puppies wouldn't be interested in salad anyway. Now rotate the state by ninety degrees, converting it into the (dogfood) state. We stick it into the box again; the puppy observes the dogfood (by smelling it, most likely) and starts barking.

But now imagine starting with the food in the (salad) state, and rotating it by 45 degrees instead of ninety degrees. We are then in an equal superposition, (food) = a(salad) + a(dogfood), with a given by one over the square root of two (about 0.71). If we were to observe it (which we won't), there would be a 50% chance (i.e., [one over the square root of two]2) that we would see salad, and a 50% chance that we would see dogfood. Now stick it into the box — what happens? If there is no puppy in there, nothing happens. If there is a puppy, we have a 50% chance that the puppy thinks it's salad and stays asleep, and a 50% chance that the puppy thinks it's dogfood and starts barking. Either way, the puppy has observed the food, and collapsed the wavefunction into either purely (salad) or purely (dogfood). So, if we don't hear any barking, either there's no puppy and the state is still in a 45-degree superposition, or there is a puppy in there and the food is in the pure (salad) state.

Let's assume that we didn't hear any barking. Next, carefully, without observing the food ourselves, take it out of the box and rotate the state by another 45 degrees. If there were no puppy in the box, all that we've done is two consecutive rotations by 45 degrees, which is simply a single rotation by 90 degrees; we've turned a pure (salad) state into a pure (dogfood) state. But if there is a puppy in there, and we didn't hear it bark, the state that emerged from the box was not a superposition, but a pure (salad) state. Our rotation therefore turns it back into the state (food) = 0.71(salad) + 0.71(dogfood). And now we observe it ourselves. If there were no puppy in the box, after all that manipulation we have a pure (dogfood) state, and we observe the food to be dogfood with probability one. But if there is a puppy inside, even in the case that we didn't hear it bark, our final observation has a (0.71)2 = 0.5 chance of finding that the food is salad! So, if we happen to go through all that work and measure the food to be salad at the end of our procedure, we can be sure there is a puppy inside the box, even though we didn't disturb it! The existence of the puppy affected the state, even though we didn't (in this branch of the wavefunction, where the puppy didn't start barking) actually interact with the puppy at all. That's "non-destructive quantum measurement," and it's the truly amazing part of this whole story.

Note that, if there were a puppy in the box in the above story, there was a 50% chance that it would start barking, despite our wishes not to disturb it. Is there any way to detect the puppy, without worrying that we might wake it up? You know there is. Start with the food again in the (salad) state. Now rotate it by just one degree, rather than by 45 degrees. That leaves the food in a state (food) = 0.999(salad) + 0.017(dogfood). [Because cos(1 degree) = 0.999 and sin(1 degree) = 0.017, if you must know.] Stick the food into the box. The chance that the puppy smells dogfood and starts barking is 0.0172 = 0.0003, a tiny number indeed. Now pull the food out, and rotate the state by another 1 degree without observing it. Stick back into the box, and repeat 90 times. If there is no puppy in there, we've just done a rotation by 90 degrees, and the food ends up in the purely (dogfood) state. If there is a puppy in there, we must accept that there is some chance of waking it up — but it's only 90*0.0003, which is less than three percent! Meanwhile, if there is a puppy in there and it doesn't bark, when we observe the final state there is a better than 97% chance that we will measure it to be (salad) — a sure sign there is a puppy inside! Thus, we have about a 95% chance of knowing for sure that there is a puppy in there, without waking it up. It's obvious enough that this procedure can, in principle, be improved as much as we like, by rotating the state by arbitrarily tiny intervals and sticking the food into the box a correspondingly large number of times. This is the "quantum Zeno effect."

So, through the miracle of quantum mechanics, we can detect whether there is a puppy in the box, even though we never disturb its state. Of course there is always some probability that we do wake it up, but by being careful we can make that probability as small as we like.

Note 2

The Schrodinger Cat riddle has been discussed by many authors, Penrose, in Penrose (2000, 2005) and in many other accounts, gives particularly detailed modern descriptions for example, and Bouwmeester (2000) gives reference to it in his book on quantum computing. Penrose (2005) gives 5 accounts (a) to (f). The Copenhagen account, (a) on Penrose (2005) page 805 clearly sounds simply a very early compromise which has worked remarkably well in some cases, the others he mentions are many-worlds, decoherence, consistent histories, split wave, some new theory (e.g. Wigner's). Equations 2.9(b) and 2.9(c) in Penrose (2000) give two accounts, for example. He defines these equations in the same semi-formal way as equation (1) above. As Penrose (2000) points out, both of these can bring problems.

In the MBI interpretation however, these problems do not arise, as we are not considering various interpretations or restatements of a known, and hopefully valid mathematically, quantum mechanical description, but a B series in which we can write the maths just as we need within the guidelines of the B series, and an A series which probably brings humans, their problems, and their past present and future, directly into play.

We know, or hope we know, that the A series exists as otherwise, along the lines of McTaggart, time is unreal. We have a century of varied philosophical and other discussions on that already.

And in the MBI A series, the human consciousness exists too. As far as we so far know, in human consciousness, outside of shamanically or drug induced cases or the like, the 'cat' will only occur in two forms, 'dead' or 'alive', and we can write this into our description of the A series as we know it. In other word the only valid human A series maps, assuming for the moment that we are considering human A series (though there is certainly no objection to considering the A series of dogs, cats, and other creatures, or indeed clever mechanical, optical or electronic devices of a deliberately or accidentally pseudo-intelligent nature) which may indeed be important in some pseudo-A or pseudo-B series, should contain no intermediate 'cat' states directly. We would just be wrong in using the wrong A series model, no paradox or riddle at all. But obviously sections of any A series bubble or indeed the entire content of it can contain by fiat no human style devices in an A series model, and then presumably Schrodinger cat states may be directly involved.. But they presumably would be involved directly in the quantum mechanics of the B series anyway, or in derived manifestations in our A series model. Of course even the point that humans cannot experience mixed 'cat' states may be disputed, looking for example of the bizarre but real mental manifestations encountered by such workers as V.S. Ramachandran or at such strange effects as we obtain in synaesthesia or in the Bonnet or Cotard syndromes, and we are considering all three but particularly synaesthesia. But if strange but surmised or actual mental effects are applicable, these can routinely be incorporated in the A series, and where necessary further work done in the B series as well.

There is clearly a lot more experimental work which can be done on the A series, to obtain models and the status of the A series more precisely. But one thing that seems clear at this stage is that the A series does exist, and can be used in a neatly understandable way to include descriptions of elaborate models involving B series results like the Schrodinger cat experiment as well.

5.04 Why did God make consciousness ? Ways and means to find out.

Introduction

"Why did God make consciousness ?" That is a question which has been asked by many people, specifically including such people as Jerry Fodor (Humphrey, 2008) who seems to have assumed, as indeed almost everyone else does, that phenomenal consciousness must be providing us with some kind of new skill. In other words, it must be helping us do something that we can do only by virtue of being conscious, in the way that, say, a bird can fly only because it has wings, or you can understand this sentence only because you know English. Humphrey thinks rather that it could be to encourage us to do something we would not do otherwise: to make us take an interest in things that otherwise would not interest us, or to mind things we otherwise would not mind, or to set ourselves goals we otherwise would not set.

The difference between these two concepts is outlined and discussed in some detail by Humphrey. Perhaps in his terms consciousness comes closer to being part of a sandbox, allowing a temporary space for the holding of various forms of reality.

Even Jerry Fodor recently claimed, "The revisions of our concepts and theories that imagining a solution will eventually require are likely to be very deep and very unsettling."

Others comment (Mindhacks, 2008) : "We need a place to erase reality and

157

redraw it or the procreative possibilities of our existence are limited by a far more slow process of biological adaptation to our environment. To experiment internally without display on the canvas of consciousness seems as impossible as experimenting in the real world without a real world. How do you test a hypothesis without positing a thesis somewhere? That somewhere is our phenomenological awareness.............I think the problem is that consciousness and unconsciousness are often thought as completely different processes, when in fact simply by focusing our attention to them we can become conscious of most of our unconsciousness processes (not all of them at once, of course, but any of them we choose to focus on). As implied in above quotes, I also think consciousness is clearly related to attention and learning new things. While adults can walks or drive a bike unconsciously and focus their attention to something else, children who are still learning these skills cannot...........It seems to me that psychologists and philosophers have come too much uncritically to accept the assumption that every biological development must serve a purpose and that all we need to do is dream some purpose and we have the explanation for the development in question. This is pretty haphazard reasoning. Evolutionary "purpose" thus becomes a substitute for Divine Will. In the case of consciousness, I don't see in what way assuming it serves a designated purpose contributes a jot to answering the putative hard question of how it arises from biophysical processes."

Recalling the 'Janus' paper by Dudai (2005) and the work of Suddendorf (1999 etc.), imagining needs use of the future and inevitably the conscious mind, thought of as part of the sandbox, must also entail the future. On the other hand, an example given by Humphrey, the triangle illusion, is one of many which do not seem to relate to the future at all, but more to a simple 'sandbox' of the mind. There could therefore be elements of truth in the views of both Fodor and Humphrey.

Now the work of Aks (2003) on self-organised criticality can almost certainly allow us to adduce a mathematical schema - or indeed several schemata - to fit the present position.

How is this related to our MBI ?

We should be able to find out by creating a test or an experiment, whether it be in psychology or in physics or indeed in both. In earlier work, experiments were done relating the 'conscious' (Juliet) mind and the 'unconscious' (Romeo) mind to impulses. These, in fact, were presented by means of a computer game which was 'Tetris' in the rough example shown, though 'Alpine Skier' may have made a stronger impulse, bearing in mind Stickgold's (Cromie, 2000) results. The experiment can be modified by altering the various parameters.

Currently, a larger series of tests of various sorts is being carried out by the Institute for Fundamental Studies in India.

As has been described earlier, one easy and direct way to do this could be rTMS. A very clear example of the strength of rTMS as a brain moulding tool is given in a simple video (Gazzaniga, 2006) and a typical example of its actual use in a somewhat cognate problem, synaesthesia, occurs in (Mattingley, 2004).

Seneker (2002) and Sagiv (2006) thus came to roughly the conclusion that neurobiological evidence shows that separate features of visual information are projected to different cortical regions of the human brain. Relatively early in the processing of visual stimuli, color and shape are separate, and the brain can encode these features without awareness. This work supports the idea of modularity in the human cortex.

It is possible that color-grapheme synaesthesia results from a flaw in the modular organization of the brain. Results agree with the possibility that cortical regions for processing shape and color are abnormally linked, but only during awareness. These findings suggested to them that attention signals associated with awareness are required to produce normal binding.

Leaving aside any theoretical problems with this idea, the practical fact is that here we have a use of brain abnormalities and TMS in order to - hypothetically at least - shed light on fundamental brain processes. Reflections on some of these ideas on brain processes have been discussed in earlier work on this website, but the point being made here is that we do have a real tool which might be able to determine quite a lot of what is going on in the brain.

There are problems here. The use of rTMS is in no way a routine procedure, can be heavily invasive, and is known to lead to brain damage in its use in certain cases. A less invasive technique is probably hypnosis, which already has given some results for Mondrians (Kosslyn, 2000),and which in my view may ultimately be of value in the field of synaesthesia. In fact Kosslyn may have come somewhere near inducing a synaesthetic effect on non synaesthetes and whilst no firm current results have yet been obtained from synaesthetes our experiments on hypnosis of colour grapheme synaesthetes suggested that hypnosis seems in our experiments to have (possibly) made an improvement to

159

the degree of synaesthetic ability, and it certainly seems to have worked at reducing alleged synaesthesia, at least temporarily, for those who desire to lose it. The latter result at least seems to agree with common sense, but we take care not to be in any way dogmatic about this.

So it may be possible to use hypnosis in the present series of experiments on the reverse Stickgold effect, since now it is felt that the 'numbing' effect (as it were) of rTMS has somewhat similar brain effects on specific brain areas, as has been illustrated by the mildly controversial work of Dudai (2008) who explains (Maxmen, 2008) "In normal memory retrieval there is a set of areas that are important so we suggest that the area in the left rostrolateral prefrontal cortex is abating this process early on, halting activity that would occur downstream."

The rTMS results seem to suggest a similar numbing, blocking or traffic jam result. I carried out few experiments in hypnosis on known/believed synaesthetes and the work looked promising. In other words, the area denies access to memory-related regions until the hypnotic cue to remember flips its switch. Now Dudai (2008) has done hypnosis experiments combined with fMRI scans to show qualitatively and quantitatively the effects on the brain during temporary imposed memory loss. It therefore may be possible to obtain results in the roughly cognate field of the reverse Stickgold effect and this may also be one way forward on the blocking effect of hypnosis.

Fortunately in the Many Bubble Interpretation, all the ideas of our Introduction above emerge as part of an explication rather than as part of a problem. We use the tentative assumption of the existence of a so-called "conscious mind" (Juliet) and an "unconscious mind" (Romeo) and allow an interplay of both. As earlier blogs have indicated, this is not the only model but it seems to be one which constellates with our overall notions. Freud, Jung and very many others (obviously including by default, conceivably even Jerry Fodor) could well think such a representation to be real enough, so, allowing for later (perhaps extreme) restrictions and modifications, it should do for us here. In Pinker's terms this is something like 'emergence'.

O'Regan (2001) pointed out that he considered that the metric quality of V1 cannot in any way be the cause for the metric quality of our experience. It is as though in order to generate letters on one's screen, the computer had to have little letters floating around in its electronics somewhere. Further he pointed out that we really have little reason to believe that dreams are pictorial. Dreamlike experiences appear to be unstable and seemingly random, though

Stickgold and ourselves have added a modicum of order to the apparent chaos. But a hallmark of dreams is this seemingly random character, particularly of detail, as where for example if there is writing on a card, it is likely to differ each time you look at the card. So there is a fair view that brain does not contain pictures of the detailed environment and even that the visual system per se lacks the resources to hold an experienced world steady. Again, with the MBI, this does not faze us, to whatever reasonable extent O'Regan's views hold.

A broader issue - Attention

Tibetan Buddhists, like various other Asian contemplatives, assert that it is possible to develop various forms of extra-sensory perception and paranormal abilities (Wallace, 1998), using attentional stability and vividness. Attention is a subject which is also currently of great interest in psychology. Some would say that attention is identical with consciousness, but others that this is not (Koch, 2006).

In the experiments of Lau (2007), using TMS, the perceived onset of intention depends at least in part on neural activity that takes place after the execution of action, which could not, in principle, have any causal impact on the action itself. An alternative view that is compatible with the data is that one function of the experience of intention might be to help clarify the ownership of actions (Wegner, 2002, 2003), which can help to guide future actions. This process could take place immediately after action execution. While Lau's conclusions need not disturb MBI enthusiasts very much, they still do seem to come from the use of TMS as a tool.

The extremely interesting article by Koch (2000) suggests how ways to manipulate attention could come closer to our efforts and to those of Stickgold (1999, 2000). Koch (2008) also insists that it is his view, amongst other things, that it is important to separate out the effects of attention from the correlates of consciousness in the search for NCC as he takes the view that it is plausible that some previously proposed NCC might have been contaminated by the neuronal correlates of attention, not consciousness. A process like their dual-task visual gymnastics seems to approximate to the sort of level of mental manoeuvres one would need to play a game of Tetris. We have already shown (and are proceeding to do further work) that such games, may have a time-reverse directional effect, in that the dreams may proceed the game playing. Obviously it would be nice to amplify this effect, amongst other possibilities.

Time Travel in Theory and Practice

The Stickgold effect was very clearly present during "Alpine skier" games, as Stickgold pointed out, and full-scale virtual reality games and tests might be expected also to show a strong effect. By full-scale virtual reality games I am thinking of games like the 'virtual switchback' games that they have at leisure resorts, which offer quite real simulation, often up to the standards of say a Link Trainer (2008). There was such a ride currently in use at an amusement arcade at Blackpool, England and it should be possible to very significantly amplify such effects in such circumstances, and to vary them much more than is possible in reality. Some attempts to set up such virtual environments have already been carried out for other purposes, for example in the different field of attempting to pursue apparent claims of extrasensory perception (Wilde, 2006).

With an environment of that kind, many further experiments could be carried out, for example experiments somewhat like those of Eagleman (2008) or even rather simpler but with added general applicability. Second Life (Physics World, 2008) could also be under consideration but as very much a 'poor man's virtual reality', it clearly has even more restricted applicability and in some ways probably lacks the scope of even games like Tetris. As well as this there are many other parameters which can be adjusted, such as variation of waking up times, times at which sleep takes place, chemical supplements and additives, light and sound stimulations and so on. Further there can be the use of monitors such as Watch-Pat and the various 'lucid dream' monitors on the market, to perhaps alert sleepers, for example as to a suitable time to wake up and record dreams.

And then of course there is every psychological trick in the book, from those employed by Stanley Milgram to tricks on false memories. For example Loftus (2005) at the University of California managed to put participants off strawberry ice cream, pickles and hard-boiled eggs by implanting false childhood memories: "In the strawberry ice cream experiment a group of students were asked to fill out forms about their food experiences and preferences. Some of the subjects were then given a computer analysis which falsely said they had become sick from eating strawberry ice cream as children. Almost 20% later agreed in a questionnaire that strawberry ice cream had made them sick and that they intended to avoid it in the future." Future studies plan to implant positive memories of fresh vegetables. (Ethics ?)

But there is much more to adjusting parameters than the above, and as we have already described, there are physical and mathematical equations on hand to do so in our proposal.

References

Aks, D.J. & Sprott, J. C. (2003) Resolving perceptual ambiguity in the Necker Cube: A dynamical systems approach. Journal of Non-linear Dynamics in Psychology & the Life Sciences, 7(2) 159-178.

Cromie W.J., (2000) Harvard Gazette Online,/2000/10.26/01, http://www.hno.harvard.edu/gazette/2000/10.26/01-sleep.html

Dudai Y., (2005),"The Janus face of Mnemosyne", Nature, Vol 434, p567

Dudai Y., (2008), "Mesmerizing Memories: Brain Substrates of Episodic Memory Suppression in Posthypnotic Amnesia", Mendelsohn A., Chalamish Y., Dudai Y., Solomonovich A., Neuron 57, 159–170, January 10

Eagleman D.M., (2008), http://neuro.bcm.edu/eagleman/

Gazzaniga M. (2006), http://www.wwnorton.com/college/psych/psychsci/media/4_tms.html from: "Psychological Science", Second Edition, Gazzaniga M., Heatherton T., ISBN-10: 0-393-92497-1 • ISBN-13: 978-0-393-92497-8, Norton Books

Hameroff S.R.,(1998), "Fundamentality: is the conscious mind subtly linked to a basic level of the universe?", Trends in Cognitive Science 2(4)119-127 ; also exchange with Spier and Thomas.

Humphrey N., (2008) "Seed Magazine" January 2008, http://www.seedmagazine.com/news/2008/01/questioning_consciousness.php

Koch C., Tsuchiya N. , (2006), "Attention and consciousness: two distinct brain processes", Trends in Cognitive Sciences Vol.11 No.1, p 16

Koch C., (2000), "How to manipulate attention" , http://www.scholarpedia.org/article/Attention_and_consciousness/how_to_ma nipulate_attention

Koch C. Tsuchiya N., (2008), "The Relationship Between Top-Down Attention and Consciousness", Toward a Science of Consciousness, April 8-12, 2008 - Tucson, Arizona

Kosslyn S. M. et al, (2000), "Hypnotic Visual Illusion Alters Color Processing in the Brain", Am J Psychiatry 157:8, August, 1279

Lau H. et al, (2007), "Manipulating the Experienced Onset of Intention after Action Execution", Journal of Cognitive Neuroscience 19:1, pp. 1–10

Link Trainer (2008), http://en.wikipedia.org/wiki/Link_Trainer ; but also see http://www.link.com/ or simply Google "Flight simulators at home".

Loftus E.F., et al (2005)," False beliefs about fattening foods can have healthy consequences", PNAS September 27, 2005 vol. 102 no. 39, 13724–13731 www.pnas.orgcgidoi10.1073pnas.0504869102

Mattingley et al (2004),Nature Neuroscience, Vol 7 (3), 217

Maxmen A., (2008), "Mind Control: Hypnosis offers amnesia clues", Science News, Jan. 12, 2008; Vol. 173, No. 2 , p. 20

Mindhacks (2008), http://www.mindhacks.com/blog/2008/01/false_trails_in_the_.html

O'Regan, J. K., Noe, A. (2001) "A sensorimotor account of vision and visual consciousness", Behavioral and Brain Sciences, 24(5)

Physics World (2008),"Doing physics in Second Life", Feb 1st, 2008 ; http://physicsworld.com/cws/article/print/32673 and sad comments at http://physicsworld.com/cws/article/print/32675;jsessionid=EB8F77783C2FB 2484BD8B16775E45E49

Sagiv, N., J. Heer & L.C. Robertson (2006), "Does binding of synesthetic color to the evoking grapheme require attention?", Cortex 42 (2): 232-242, PMID 16683497

Seneker S.S., (2002) M.A. thesis, East Tennessee State University

Stickgold, R., Malia, A. & Hobson, J.A. (1999) "Sleep onset memory reprocessing and Tetris. Journal of Cognitive Neuroscience" 11(supplement)

Stickgold, R., et al , (2000), "Replaying the Game: Hypnagogic Images in Normals and Amnesics" Science 290 (5490), 350. [DOI: 10.1126/science.290.5490.350]

Strogatz S.H., Mirollo R., (2007), "The Spectrum of the Partially Locked State for the Kuramoto Model", Journal of Nonlinear Science, Volume 17, Number 4, August, 2007, 309-347 (online)

Suddendorf, T. (1999). The rise of the metamind. In M.C. Corballis and S. Lea (Eds.), The descent of mind: Psychological perspectives on hominid evolution (pp. 218-260). London: Oxford University Press

Wallace B.A.,(1998), "Training the attention and exploring consciousness in Tibetan buddhism", "Toward a Science of Consciousness", Abstract No:748, "Tucson III" April 27-May 2, 1998 The University of Arizona Tucson, Arizona

Wegner, D. M. (2002), "The illusion of conscious will", Cambridge: MIT Press.

Wegner, D. M. (2003), "The mind's best trick: How we experience conscious will". Trends in Cognitive Sciences,7, 65–69.

Wilde, D.J et al (2006), "The design and implementation of the Telepathic Immersive Virtual Reality System", Paper Presented at the 2006 Parapsychological Association Convention, Stockholm,Sweden.

Zhaoping L., (2006), "Theoretical understanding of the early visual processes by data compression and data selection", Network: computation in neural systems, December 2006, Vol 17, Number 4, Page 301-334

Zhaoping L., (2007), " A bottom up visual saliency map in the primary visual cortex --- theory and its experimental tests", Adapted from the invited presentation at COSYNE (Computational and Systems Neuroscience) conference, Salt Lake City, Utah, USA. Feb. 2007
http://www.cs.ucl.ac.uk/staff/Z.Li/prints/CosyneForPost.pdf

Zhaoping L., (2002), " Computational vision --- a window to our brain", Presentation CUSPEA conference on "Physics in twenty first century" in June 2002, Beijing. http://www.cs.ucl.ac.uk/staff/Z.Li/prints/talk.ppt

Zhaoping L., (2008), ScienceDaily, Feb. 19, 2008
http://www.sciencedaily.com/releases/2008/02/080215103210.htm

Time Travel in Theory and Practice

Appendix: Computational neuroscience and some of its problems

Computational neuroscience as such, is an important subject insofar as it places itself to ask many awkward questions, and often expresses great dissatisfaction with some of the answers. Some of these difficult processes are often applicable in other contexts as well. I give the example of Zhaoping (2008) who recently apparently pointed out "When you see them throw a ball into the air, followed by a second ball, and then a third ball which 'magically' disappears, you wonder how they did it. In truth, there's often no third ball - it's just our brain being deceived by the context, telling us that we really did see three balls launched into the air, one after the other. .... Mathematical modelling suggests that visual inference through context is processed in the brain beyond the primary visual cortex. By starting with a relatively simple experiment such as this, where visual input can be more easily and systematically manipulated, we are gaining a better understanding of how context influences what we see. Further studies along these lines can hopefully enable us to dissect the workings behind more complex and wondrous illusions". This simple sort of experiment, when redefined in terms of the neuroscience lab rather than those of the juggler, certainly looks like a excellent way to relate context to details and has a close correspondence with other more detailed lucubrations, well illustrated in Zhaoping (2002, 2006, 2007).

But we are not obliged to accept all of Zhaoping (2002, 2006, 2007) or indeed any of it, while we might still retain the brute 'common sense' of the idea about the three balls. Clearly we may accept some of Zhaoping's other reasoning in Zhaoping (2002, 2006, 2007) - and almost certainly might do so - but it is a far step to say we accept it all, or indeed accept in its entirety some such line of research in computational neuroscience. These are topics that philosophers argue about, each perhaps convinced that they are right. In our situation, with the MBI, we could stake the claim that we have ideas that are founded at the very highest level, that of McTaggart's paradox and large category theory - and then go ahead and use a great deal of computational neuroscience to further our work.

Anything like computational neuroscience must be 'down' from McTaggart's paradox and large category theory, we could argue, just as we could argue that simple sums are 'down' from Peano algebras though doubtless the whole philosophical idea of a hierarchy of thought could be quibbled at - but it is probably easier to look at specific facts in computational neuroscience, for example the possibility of overtraining. Now there are certainly checks which

166

can be placed on that process, as there are several methods available to check the degree of generalization and/or to detect overfitting, for example cross-validation and the use of noise addition - and it is not suggested that Zhaoping has overtrained anyway. But the fact still remains that a lot of such 'small' factors could arise in an intricate scheme, and one way or another they usually do, at some point.

An example of where such a thing can admittedly happen is suggested through the work of Strogatz (2007) for example, where the elegant work of Kuramoto seems to have led to rather difficult complex system theory, even though Kuramoto's actual results seem to serve for useful models in many situations. Many attempted 'expansions' of Kuramoto's approach seem to go no further, for various good reasons. On the other hand the ideas of McTaggart's paradox and large categories stand well above all this, and are not specific in nature, even to the extent for example that we are not simply having a disagreement over the relative merits of neutral monism for example, which is about the sort of level to which argumentation over the Penrose/Hameroff (Hameroff, 1998) ideas at one time descended. The real crux of the Penrose/Hameroff matter probably was that Penrose/Hameroff was wrong about microtubules, superconductivity etc. etc. but right that the Eccles-Popper overall viewpoint of Penrose/Hameroff could not be simply thrown over readily by (basically naive) computational neuroscience, whether rightly or wrongly, for a partial reason that it was defined and delineated in roughly Popper-Eccles terms. That level of 'philosophical' dispute is not what we are looking for.

Clearly then, what we must do is to use computational neuroscience to help our measurements but not accept it as necessity in any way in our descriptions. In other words we must use the results of computational neuroscience with reason, if we can relate them in a useful way to other claims. The Zhaoping example above, therefore, looks like we could use it somewhere in an experiment fairly readily but that does not mean that its context must, in essence, ruin other assumptions or views which we might have, or might wish wish to consider later.

6

Developing the Theory

In Chapter 6, Sections 6.01 and 6.02 proceed to develop a cross-over between the A and B series. Section 6.03 relates strange attractors, chaos theory and the Wason test to both A and B series and the duck/rabbit illusion. Section 6.04 brings in emotion and feelings to the descriptions, as well as free will, physiological drives and the like. Section 6.05 suggests ways that some mental manifestations may be managed, in particularly less likely ones. Section 6.06 includes various applications of the sandpile model to the MBI, the pseudo-A series, and more ideas on the Reverse Stickgold effect.

6.01 A new cross-over from McTaggart A-series to B-series

Abstract

An attempt is made to find sources of new parameters whereby an essentially B-series model of A-series matters can be more accurately obtained. These matters will, it is hoped, allow specific physical parameters to be applied to a study of human consciousness. To do this, the experiment of Libet (1985) has had to be re-examined.

169

Time Travel in Theory and Practice

Introduction

A new - quite possibly the first - clear cross-over from McTaggart A-series to B-series is being considered. This may give some new parameters to work with in the studies of consciousness.

The large amount of dreamwork available today claims to refer mainly to effects in the part of the mind considered during dreams (Yates, 2009a etc) and dream results have traditionally seemed inchoate, often contradictory and hard to fathom. Nonetheless they are observable phenomena and should therefore be regarded as such. Stickgold and others have found correlations between dreams and waking states so a full description of "consciousness" should involve them.

In the present note,we go on to consider what could be regarded as more tangible or concrete results, namely regarding the Libet experiment or so-called "Libet half-second" matter.

Work relating to the Libet (1985) experiment has often traditionally been regarded as relating directly to the 'conscious person', such as he may be. This is because in this experiment, individuals have to determine or estimate mentally when they decided on a particular task. The recent work of Banks (2008), though not necessarily being used exactly in the way Banks (2009) intended, can help to show how these results might actually be used. However we care to look at it, these results are interesting in my opinion.

First a brief survey of some aspects of the Libet experiment is perhaps due, as any parameters referring directly to the mind may be of use in these lucubrations (Yates, 2009).

Brief survey of some aspects of the Libet experiment

In the important blog "Conscious Entities" (2009) the question has been posed "Libet was wrong ?" and that blog suggests that in some respects Libet may have been wrong over the "Libet half- second" matter. Largely it is suggested and explained briefly but fully, that the work of Trevena, mentioned in this blog (Yates, 2009) and referred to in "Conscious Entities" (2009) is significant in that regard. In effect the EMG measurements are just manifestations of neural activity and provide no totally complete neural markers. But in considering Trevena, the free will enthusiasts are trying to recover the free will concept within the B-series. It seems to me that to try to recover the free will concept inside a block time model is a far fetched idea.

The block time model was developed on the basis of the calculus of Newton and Leibniz who both seemed to believe (or to pretend to believe) in some sort of all-powerful God, and whilst their beliefs ran in somewhat different directions, most of modern physics (including general relativity and quantum mechanics) seem to include an essential ingredient which gives us this block time model (however varied and tortured), without the possibility of free will in the sense accepted in what is now termed 'folk psychology' and is the subject of (often very desultory) surveys by the X-phi community. Block time is more like a map of a country showing say "past" as the south and "future" as the north with effectively only one way traffic from South to North. As Feynman (1970) for example illustrated, there may well be nothing wrong with that one way street. Nonetheless - even if it is not a bug - the one way street is only a B-series feature. It is not necessarily an A-series feature, in fact it probably isn't. And that is independent of whether most proposed B-series time travel supposed thought experiments - of which by now there are many, usually involving wormholes and the like - actually work in practice.

Anyway the Trevena tests should have been done long ago. For the moment we can possibly just assume from them that EMG evidence for an earlier unconscious intention is not supplied in enough detail to make totally adequate assumptions from.

Now referring to Bank's work. Banks has of course provided us with a detailed book (Pockett, 2006) which contains much work on the human aspect of the matter. The more recent work (Banks, 2008, 2009) implies considerable variations in the 'decision time' W.

In private communication with us Banks mentioned the following "Meanwhile, it's not the case that W is always about -200 ms. The values in the literature range from about -100 ms to -1.42 seconds by Matsuhashi & Hallett, 2008. Soon, et al (Soon , C.S., Brass, M., Heinze, H., & Haynes, J. 2008. Unconscious determinants of free decisions in the human brain. Nature Neuroscience, 11, 543-545) found the RP to begin very early (up to 7 seconds before the response), and the estimated W inn their study to be about -600 ms. My article with Isham makes a qualitative point that should not be confused in any way with the precise number of milliseconds W is shifted by the deceptive auditory cue. The point is that W is affected by an event that comes after the response. This finding is evidence that the W people report is a retrospective inference from their observation of when they respond. I am writing up a more elaborated theory of the response and the estimate of W that I can send when it is ready. The point is that the action being judged for W in the Libet paradigm is at the level of intention-in- action (in Searle's terms), and it has been recognized at least since Lashley that we have no conscious access to our behavior at this level."

Now it remains to be seen that W is actually obtained from 'retrospective inference' within the B- series. If Banks is correct as I assume pro tem, it is very likely to be considered as such.

I would have expected on the face of it that W should be about equal to the human reaction time in the B-series if decisions were intended to be followed immediately by keypresses. In the Libet experiment when actually carried out in the way that it usually is, the subject seems to be rather dragooned into pressing the key right way, and this is almost an experimenter's command. The whole matter needs to be a lot more thought out, "First awareness of a wish to act" are the specific words used by Libet in his 1985 experiment as apparently part of a definition of W, which places W clearly as a marker which should at the very least be within the bailiwick of any consciousness theory and more specifically could be an actual measurement of an A-series result. (One is reminded that the A-series does have 'actual' past, present, and future by definition and may not be one to one mappable to a B-series time model).

Conclusion

The variations in W and also in any timings concerning the keypress we hope to be able to investigate for a number of subjects over a period of time, also using various audiovisual distractions, possibly including some of a so-called 'subliminal' nature, bearing in mind such matters as the work of Phil Merikle and of Nilli Lavie. We also do not know that the B-series is not fundamentally flawed per se . So a good A-series model (even in crude B-series representation) may also be better than a simple traditional B-series model, though there are no current hopes in that regard.

We also hoped that the late Professor Banks would have been able to describe his experiments and his own views in detail at our conference next year, as he had kindly agreed to do.

References

Banks W.P., Isham E.A., (2008) "We Infer Rather Than Perceive the Moment We Decided to Act", Psychological Science,Vol 20, Issue 1, Pages 17 - 21

Banks W.P., Isham E.A., (2009) "Do we really know what we are doing? Implications of reported time of decision for theories of volition". In: Nadel L., Sinnott-Armstrong W. P.. "Conscious Will and Responsibility: A Tribute to Benjamin Libet". Oxford University Press, in press

Conscious Entities (2009) September 26, 2009, "Libet was wrong ?" , http://www.consciousentities.com/?p=233

Feynman R.P., (1970), "Feynman lectures in Physics", especially near end of chapter on Entropy.

Libet B., (1985), "Unconscious Cerebral Initiative and the Role of Conscious Will in Voluntary Action." Behavioral and Brain Sciences 8: 529-66 ; enormous amount of other work such as Libet, B. 2004. Mind Time. Cambridge, Mass. Harvard University Press.

Pockett S., Banks W.P., Gallagher S., (2006), "Does Consciousness Cause Behavior?", MIT Press, ISBN: 978-0-262-16237-1

Yates J., (2009) http://ttjohn.blogspot.com/2009/09/new-look-at-fitzhugh-nagumo-method- in.html#links

Yates J., (2009a), http://ttjohn.blogspot.com/2009/05/many-bubble-interpretation- externalism.html#links , http://philpapers.org/archive/YATTMB

6.02 A new look at the Fitzhugh-Nagumo method in McTaggart A- series simulation, together with the use of solitons or chaos theory

In our dynamical systems models (1) for the waking and sleeping brain, we used Berkeley Madonna for the simulations and after exploring a very wide number of possibilities and many parameter values. The Fitzhugh-Nagumo

(FHN) model is discussed briefly in Section E of that paper. The conclusion was come to in that paper that "In common with results for many cases where modelling is made slightly more complicated but requires more parameters, so far (the FHN) does not seem to have really paid off at this level of model making. It might be a way forward at a later date however." However some quite satisfactory results were obtained with slightly simpler models, such as the one referred to as N003b.

I had in mind numerous previous instances such as the great initial effectiveness of the Kuramoto model, its obvious applicability to many systems and yet the extremely difficult process of refining it much further in specific cases. The Kuramoto model seems to have been useful in a general way in areas as varied as descriptions of neural processes and the London Millennium Bridge.

Solitons

Since we are working in reference (1) in the region of models like the FHN model it seemed reasonable to consider whether any alternative or sufficiently differing brain models might produce better results. The Soliton Model (2) in neuroscience is justified as follows: "The model starts with the observation that cell membranes always have a freezing point (the temperature below which the consistency changes from fluid to gel-like) only slightly below the organism's body temperature, and this allows for the propagation of solitons. It has been known for several decades that an action potential travelling along a neuron results in a slight increase in temperature followed by a decrease in temperature. The decrease is not explained by the Hodgkin-Huxley model (electrical charges travelling through a resistor always produce heat), but travelling solitons do not lose energy in this way and the observed temperature profile is consistent with the Soliton model. Further, it has been observed that a signal travelling along a neuron results in a slight local thickening of the membrane and a force acting outwards; this effect is not explained by the Hodgkin-Huxley model but is clearly consistent with the Soliton model. It is undeniable that an electrical signal can be observed when an action potential propagates along a neuron. The Soliton model explains this as follows: the travelling soliton locally changes density and thickness of the membrane, and since the membrane contains many charged and polar substances, this will result in an electrical effect, akin to piezoelectricity."

Now this is a new model and differs in very significant ways from the FHN model and similar models, and whilst it is claimed to have many advantages, such as in an understanding of the Meyer-Overton observation, this attempt to explain nerve transmission by sound impulses rather than simply electrical impulses certainly has not replaced the conventional model. At the same time, the model we were trying to use is basically an interpretation of the A series using B series mathematics. It is only a model not an elixir, and the position is very like that of the traditional John Godfrey Saxe description of the blind man describing an elephant - it will not be right in every detail. David Corfield's general suggestion (3) involving the use of vector solitons, as has been already used in somewhat similar cases, could well be a further way forward.

And this is not a walk in complete darkness - consciousness has frequently been described as an emergent phenomenon in a collection of neurons, as indeed have matter wave solitons and optical solitons been described as emergent and placed in the same category. Filamentation is a related phenomenon as exemplified by meandering rivers and lightning bolts.

And, just as consciousness can clearly be said to presumably relate to the P=NP? problem, in 2002 I had considered looking at soliton theory and the Backlund transformation, in the hope that Mielnik's idea could be extended.

Further, it has even been suggested by Hameroff and Penrose (4), that quantum computation in the brain works by solitons. Both Hameroff and Penrose have produced many interesting ideas, though this one has encountered much opposition, so it is mentioned here although we do not propose to use it at this time.

So as distinct from simple philosophical argumentation and questioning - still important tools - we can do the calculations in Berkeley Madonna, without - and this is a key and important point - losing important philosophical stringency in the way that seemingly began in quantum physics on the introduction of the Copenhagen Interpretation and then got to the point where a dog can now seem to be able to understand quantum mechanics better than a human can.

But we already have the added problem in X-phi (experimental philosophy) of a trend to mathematical oversimplification and a rush to philosophical relativism almost like a Hollywood star might run to a Dr. Feelgood with dire consequences, so a lot more work needs to be done steadily and carefully in X-phi also.

175

I do not yet know if the soliton approximation will help, but it is a matter of trying it for various cases without seeking a mental Theory of Everything, and solitons could be said to be more physically realistic than FHN though for the moment model N003b is still the top priority.

Chaos Theory

Another important model is described in a video (5). This can possibly correspond to the sandpile effect we have been mentioning in this blog for some time now.

The effect is now discussed in the eminent and well recognised Conscious Entities blog (6), which particularly states "One claim not made in the article, but one which could well be made, is that all this might account for the sensation of free will."

I have to agree with that possibility and reference (1) of course remains open to that effect and indeed briefly discusses it as the "butterfly effect". You would expect a chaos effect to arise in any model which allows for the so-called 'unconscious mind' as its existence is what might be called a 'brute fact' as clearly the conscious mind is not capable at this time of fathoming the hidden realms of day to day consciousness. Hence there is scope in our present models for both dreams and chaos. Furthermore there is certainly no immediate requirement for 'pure chance' or 'god' or some sort of 'blind watchmaker' or indeed a 'homunculus', because of where our theory has come from.

Conclusions

So the next step is possibly to consider the recent work of Banks (7), which may have produced a live psychological experiment, not involving brain tampering, which provides a clear physical example of the Libet and Haynes effects.

Provided we realise that any instances of our brain model are merely partial mappings of the A series to the B series, there should be no conflict with free will concepts, and further progress may be becoming clearer.

176

References

1. http://ttjohn.blogspot.com/2007/05/work-in-progress-on-application-of.html#links

2. http://en.wikipedia.org/wiki/Soliton_model

3. Corfield D., (2009)
http://golem.ph.utexas.edu/category/2009/09/where_have_all_the_solitons_go.html#c026247

4. Hagan S., Hameroff S.R., Tuszynski J.A., (2000), Decoherence and Biological Feasibility, arXiv:quant-ph/0005025v1

5. Video, in "Disorderly genius: How chaos drives the brain", New Scientist, 29 June 2009,
http://brightcove.vo.llnwd.net/d7/unsecured/media/981571807/981571807_27451004001_chaotic -brain.flv?
videoId=27532501001&lineUpId=&pubId=981571807&playerId=1873822884&playerTag=&affiliat eId=

6. http://www.consciousentities.com/?p=202

7. Banks W.P., Isham E.A., (2009), Psychol Sci. 2009 Jan;20(1):17-21,
http://www.ncbi.nlm.nih.gov/pubmed/19152537

6.03 Are Physicists suffering from the Tychonic Illusion ?

To begin with we may examine the duckrabbit picture

Along with such figures as the Necker cube and the Schroeder staircase, Jastrow used the duck-rabbit to make the point that perception is not just a product of the stimulus, but also of mental activity – that we see with the mind as well as the eye. From a constructivist point of view, many illusions illustrate the role of unconscious inferences in perception, while the ambiguous figures illustrate the role of expectations, world-knowledge, and the direction of attention (Long and Toppino, 2004).

For example, children tested on Easter Sunday are more likely to see the figure as a rabbit; if tested on a Sunday in October, they tend to see it as a duck or similar bird (Brugger and Brugger, 1993).

Kihlstrom (2004) and many others have written at great length on this and related perceptual phenomena, and of course many and varied views have been expressed.

And this ties in (Pearson, 1998 ; Margolis, 1998) with Margolis's work, which I have frequently referred to in this blog, on the Tychonic illusion. There was a long correspondence in the literature (Psycholoquy, 1998) but the upshot seems to be that the Tychonic illusion subsists, (Margolis, 2002a) whatever the worldly circumstances of Tycho at the time. I discussed all this in several places before, in my blog on Wed, 17 Aug 2005, for example. I state a portion of this article on Margolis's book below for convenience (Note 1).

In actual fact the matter also ties in with the Wason selection test, which I'll assume is generally well known. On this, Margolis (2002) says "So are cognitive illusions likely to occur and be hard to correct beyond the narrow contexts of psychology experiments? I think that is inevitable, so that it is important that we come to understand these illusions. Careful thought about simple puzzles like Wason can help illuminate what happens in vastly more consequential realms."

Now in both the present circumstances and those of Tycho the position and status of the Tychonic illusion, to those not duped by it, is fairly clear.

In Tycho's case it surrounds the controversy (involving the Pope and various nobles, scientists and others) concerning the disposition of the solar system, now generally clarified, we hope. Margolis and others have already discussed that. To put it crudely, it is almost as if the solar system were just a large duckrabbit, unclearly unidentified or a Wason card test unsolved.

In the present case we are dealing with the properties of time and how it is observed by us. The simplistic way, as adopted by Einstein, the quantum theorists and the computationalists, is to go right ahead and assume we have a block universe (a McTaggart B series) and hang it all on that. (Note 3).

To pursue the analogy a little further, it may be all very well to hang a caravan on the back of a car, by contrivance we can even include the kitchen sink in the caravan hung on the back, but it would be extraordinary folly to hang a 4 bedroom detached house on the back of the car. I will accept that a house move may be possible in that way (I think using heavy trucks it has even been done in Vancouver Island) but it is not a practical way to run things generally and would lead to stress and hardship. In the same way, by its nature the B series does not contain a past, a present and a future, it does not distinguish in any acceptable way any real element of freewill, and plainly does not represent the universe as we know it. Einstein may well have said "God does not play dice" but it is torturing facts to say that "God created a Universe without freewill, he created humans as robots controlled as by a computer, and he laid it all out for us to see". That God does not correspond to the God of Einstein's somewhat dodgy anthropomorphisms or to that of anyone but a scientist who has made too many assumptions and believes in no grounds to justify them. Or if you like, has used an oversimple model. If we say all that, where is God or even further, where is the possibility of conceiving that there may be a God. Certainly the B series is a useful working tool, but it is hardly even a full hypothesis.

It is unscientific to assume that God will or could somehow appear in the works when we know that possibility is right there from the start. The possibility of God is a fundamental axiom of any physical description of the universe and we know that simply because we know that we have that possibility before we start. To say otherwise is like assuming that some important factor will eventually cancel out from our sums, and therefore leaving it out from the start. That is clearly unscientific. Obviously there are ways of leaving out irrelevant factors before we do start. Traditionally, atheists try to do this. My experience is that a convinced atheist is the only person I can think of more philosophically frightening than a convinced cleric, be he Christian, Muslim, or any other. Agnosticism still leaves the door open for God and allows God into a theory as someone's idea, however outrageous it may seem to some. Now given the possibility of God, we do not have to assume he is just or fair, but we could assume that we are rational so it is reasonable to allow the further 'working assumption' that God, if he is indeed there, may be rational too. But the B series rationality of a God escapes most people especially those who see the Tychonic illusion and even more, Wason selection.

Fortunately we have the A series as well, which allows past, present and future to individuals, and the possibility of free will, known to us all in some sense. We don't have to have a past, a present and a future or freewill of some kind if there is an A series. But if we believe our senses we are living with a past, a present and a future and hopefully with some element of freewill - indeed we may be optimistic enough to believe in the existence of a higher power (or powers), called God (or Gods).

As far as I am aware an easy mathematical representation of what is happening in the A series may sometimes be made in terms of some ordinary mathematical techniques already familiar to us from their use in the B series, and that of course there is a tendency to do, but that of course cannot be allowed to confuse the issue of a fundamental difference between the A series and the B series. And it is necessary to remember what we are doing, and that any mathematics which we write down as supposedly representing happenings in the A series is not necessarily an identical representation of what is going on in the A series, and it certainly would not transfer into a piece of B series mathematics without careful thought, if at all, and indeed the mathematics may at best only be a partial representation of what is going on in the A series and certainly not necessarily a complete model of some part of the A series nor representing all the sufficient and necessary constraints of that part.

Certainly, most people would see from common observation that we are not "simply" like in the B series, just in effect a lot of flies preserved for ever in some timeless aspic or solid jelly. And I certainly do not take a Panglossian view that all that will "eventually" come out of the existing physical theories. Yes indeed, it seems that modern physics tends not only to be caught up in the

Tychonic illusion but simply to be Panglossian about it. Certainly physics, and even we may hope the future of say the Higgs boson, may have achieved remarkable results during the last 100 years and we may hope that it continues to do so. But we cannot assume that for that reason it has comfortably set out its mathematical stall to adequately cover important and relevant fields like human perception, consciousness and insight, even insofar as these matters are directly relevant to physics. I think of the early expositions of special relativity in this context, and see how far physics has proceeded in just using them, and their simple analogies of time and distance ! How much further can it proceed using the A series of McTaggart. Perhaps modern string theory (Greene, 1999 and many others pro, con and doubtful) could eventually get us out of our problem, but that in a way, could be just like the Ptolemaic epicycles getting us out of the idea of the earth revolving around the sun. The symptoms looks similar and the matter seems highly Tychonic so starting with an A series and a B series seems the right way around to go about things. It will not damage modern physics as that is in the B series already, nor will it force us to adopt some bizarre new theory. (Note 2) Instead it will complement and add to existing physics, as well as being based on the ideas of someone who was undoubtedly one of Britain's greatest philosophers, namely J.M.E. McTaggart.

It seems to me that one reason why we are in the grim position that we are, may be the dogged insistence of atheists (often wrongly trying to describe themselves as agnostics) trying to leave out anything remotely Godly (almost in an HG Wellsian way) set up against the often equally dogged insistence of clerics to presume some strange opposite view, often enough without even trying to add reason to that view. (I'm sure we have all heard the American snake oil evangelists who often do not have even any decent snake oil). Each party seems to be trying to drive the other to strange excesses.

Clearly if a person uses the A series as well as the B series it does not automatically allow the assumption that such a person has a belief in God, but it does imply perhaps that he knows that there are other people, probably fairly rational in some cases, who do believe in a God or Gods. In fact one could possibly deduce that for a person not to have an A series at all is possibly for him to commit a philosophical solipsism to the point that arguments made by such a person are philosophically incorrect. (Rather like the conference of solipsists said to be the butt of jokes in that their organiser was so pleased to see so many people of the same mind). Even leaving aside the God question, a similarly convincing argument can possibly be made in terms of human psychology but it might be a little longer.

The question of using chaos theory and strange attractors is another issue which I am also exploring at present (Yates, 2006a, 2006b, 2006c, 2006d, 2006e, 2006f, 2006g, 2007a). And of course I bear in mind the work of Freeman. Especially in considering Freeman's work, we must also give careful attention to the old maxim "the map is not the country" but given that, there is, for example, a very simple suggestion from Calvin (1996) to consider in that context: "When the Necker cube switches back and forth between top-down and bottom-up perspectives, it's presumably because we're switching in and out of lobes of an attractor". Now this can be fitted in with the A series in numerous ways but it is not immediately philosophically evident that the "flies in aspic" B series can properly philosophically cope with it at all, except perhaps in some Panglossian way.

References

Brugger, P., & Brugger, S. (1993). The Easter Bunny in October: Is it disguised as a duck? Perceptual & Motor Skills, 76, 577-578.

Calvin W., (1996), "The Cerebral Code", Chapter 5, MIT Press, ISBN 0262531542 ;
http://williamcalvin.com/bk9/bk9ch5.htm

Greene B., (1999), "The Elegant Universe", ISBN 0393 058581

Jastrow, J. (1899). The mind's eye. Popular Science Monthly, 54, 299-312.

Kihlstrom J.F, (2004) http://ist-socrates.berkeley.edu/~kihlstrm/JastrowDuck.htm

Long, G.M., Toppino, T.C. (2004). Enduring interest in perceptual ambiguity: Alternating views of reversible figures. Psychological Bulletin, 130, 748-768.

Margolis H., (1998) "Tychonic Illusions: Hard vs. Easy". Psycoloquy: 9(38)

Margolis H., (2002), "Wason's Puzzle and Real Problems", AIBS conference on behavioral economics, Great Barrington MA, 2002)

Margolis H., (2002a), "It started with Copernicus", McGraw-Hill. ISBN 007 138507X

Pearson D., (1998) "Imagery need not be blind to fail - Commentary on

Margolis on Cognitive- Illusion" Psycoloquy: 9(34) Psycoloquy topic
Cognitive Illusion, (1998) , http://www.cogsci.ecs.soton.ac.uk/cgi/psyc/ptopic?
topic=Cognitive-illusion

Yates J. (2007a), Memetics and the A Series (3); references to the rest mainly
in earlier blogs.

Notes

1. Margolis "It Started with Copernicus": This book begins with very
ambitious claims. Margolis gives a very impressive Table I-1, giving a list of
scientific discoveries made around the year 1600 and a further (empty,
qualified) list of the work done in their respect in the previous 14 centuries. He
then states on page 6 " The sharp step that is visible about 1600 almost by
definition requires something new" and he says that this was not merely
improved experimentation or indeed new mathematical techniques.

Margolis is a distinguished writer so these claims must be taken seriously and
considered in conjunction with his work on Wason tests.

We have to try to work out 'what 'something new ' was involved, and why.

Now the rest of the book is mainly a historical exposition and we have to sift
through the history to find explanations of current relevance.

Firstly one has to mention that he really may have something new to say,
bearing in mind for example his Wason conclusions and also his historical
comments about Tycho. The latter are summarised on p48 "It would be 400
years before anyone noticed" (the Tychonic illusion). Following this matter up
later he says."we have also seen that many logically accessible discoveries
waited 2000 years to be made" (p200) and compare Wolpert "... find a new
phenomenon where you could open up a whole new world . you'd jump on it
like a shot" ([q-ball] The Telepathy Debate, 18/07/05).

Now what Wolpert says is unfortunate but common and easily sympathised
with. Margolis's idea really ought to help to sort it out. Of course we must bear
in mind the academic discussions (for example by Topper) on Margolis's
Tycho conjecture but as far as I can see, Margolis has a case to proceed with
his views - and indeed he does proceed. I can think of several other problems
with his comments on Tycho, the chief two being that I am far from sure that
Margolis was the first to notice this matter as I had heard it elsewhere much

previously but haven't tried to locate it in the scholarly tomes - and secondly I doubt that many people cared all that much for the last 300 or so years. But these points may matter little or more likely, may fit in with Margolis's own theory, when expounded in detail, if and when it can be.

Also, the Tychonic illusion is likely to have been one illusion amongst many.

On page 26 he criticises Kuhn for use of the term "aesthetic" to describe a part of the valid approach towards theory-selection [At this time I must make the point Kuhn and Popper schools of thought are often regarded as different and Americans often tend to regard themselves as Kuhnians; to me, Popper and Kuhn both appear to be thoughtful philosophers of their time]. Margolis however seems to prefer the word "economy" to deal with what he is talking about and speaks of a "cognitive fit" to make an idea "comfortable". In terms of modern science this idea can be perhaps be used more readily with brain-scanning, though Zeki has spent a considerable time on aesthetic points of view towards brain scans too.

But Margolis points out that an uncomfortable fit of ideas can improve with time, in the sense that 'everyone believes it' and this makes it more comfortable. This probably does not apply to a case like the aesthetic value of Mozart as compared to Salieri. which is unlikely to change significantly with time. So it looks to me as if we already have a way to distinguish between the "aesthetic" and the "economical" with regards to brain scans. One area will change with time and probably circumstance and one area will not, and the cases can be shown on a graph. Indeed we can probably dig out brain scan ideas for 'comfortable' as well.

Most people could already construct a theory comparing and contrasting say "aha!" to "aesthetic", and bear in mind that acclimatisation can occur quite quickly.

Margolis makes on p56 the important point that the difference between a Ptolemaic system and a Tychonic system was not AT THAT TIME (my emphasised capitals) logical. Indeed it was aesthetic (or maybe economic). As Kepler pointed out, God could have set up the solar system in pretzel shapes had he wanted. But eventually comfort drove people to Tycho and those after him. Easier to do circles than pretzels in effect.

Now at this point we are clearly tempted to try to work out if such a principle can reasonably be generalised, to the work of say Turing and indeed

Schmidhuber and others. Well the first answer is, that such authors will probably automatically try to simplify their models either through their own mathematics or on some metamathematical principle, such as Godel's theorem which in Godel's original formulation was very long but, if you have the patience, very comprehensible. Thus we can certainly simplify models, in size or complexity or understandability, and do our best to choose the simplest one, but the 'beauty' or 'aesthetics' or indeed 'economy' or 'comfort' of an idea will probably need more MRI scans to obtain correlation coefficients. Anyway we still have problems with the white matter, the fact that most results are MRI not DT-MRI and fundamental facts such that to use simple blood flow in this way almost smacks of phrenology.

Anyway there is at least a brisk platform, we now have a pulpit to preach from as they might say in adspeak.

Margolis noted (p58) that Tycho's work was an enormous wrench in terms of comfort away from Ptolemy, even though he still took the sun to travel round the earth. He destroyed the Ptolemaic idea of the spheres and moved the orbits of Mercury and Venus, and this fitted facts as then known. He got away with it possibly because by then Copernicus's work was somewhat accepted or at least scientifically recognised as a theory. It is arguable that Tycho was by now just climbing on to a bandwagon, but simply had respect for Rome. It is almost like the end of Marxism, but it is unlike to be like the decline of capitalism as capitalism has tried to make sure that no workable alternative system remains. A faith like Islam is unlikely to fill the gap capitalism will leave. As they say, you cannot beat the system (or Bilderberg, it seems).

So at least Margolis implies a recipe: when a good new theory comes, take away the unobjectionable parts of the old theory, replace them with as much of the new approach as possible, and leave the kingpin of the old theory till last, like gradually and competently removing the foundations of a house. Very Machiavellian, or perhaps purblind and naive?

Margolis says, however, 'why should a (partly) heliocentric system (Tychonic) be appealing to a geocentric astronomer'? Margolis refers to his two qualities 'economy' and 'comfort'. He then refers to Zajonc's experiments, which roughly involved subliminal presentations being supposed to enforce liking partly through familiarity (In the spirit of 'Mother's cooking is best' i.e. repeated exposure to a stimulus brings about an attitude change in relation to the stimulus). It's probably worth noting that Zajonc's work tended to imply reinforcement rather than liking. Anyway Copernicus's view and Tycho's view became familiar if not always accepted. But the only clear advantage that Tycho's view had was that it was similar to Copernicus's view (p64)! One can reach the conclusion, if one chooses, that free speech without penalty could have allowed Copernicus's view to predominate earlier. But Margolis stresses continuing comfort as being important, Tycho's view being substantially similar to Copernicus's. Easier for comfort, to overlook the errors in Tycho's work.

Basically we are left almost with an Edison/Tesla situation. Edison gave Tesla a great contract for the use of Tesla's alternating current (as opposed to Edison's inferior DC) reticulation - the electric chair for executions. In the same way, we could reckon Tycho left the heliocentric solar system idea to Copernicans.

So we have a rough blueprint for the use of MRI for discovery. Aesthetics, economy, comfort, MRI, and correlation coefficients. This lays the groundwork, I may write later in more detail.

2. It is probably unfair to call Penrose's twistor theory a bizarre new theory as I remember first glancing through the proofs of a version of Penrose's twistor theory in Sale, Cheshire, nearly 40 years ago. By now it is a bizarre old theory I suppose, but in its present form it has many competing theories to contend with. The fact remains that even today, the future of string and gravitational theory, and certainly of the TOEs is nowhere near settled.

3. Nearly all decent philosophy conferences dealing with time these days devote a lot of conjecture to the ideas of tensed and tenseless time, which still are very much on the philosophy agenda (as indeed I have mentioned in more detail in earlier blogs of this series) even if not clearly on the physics agenda.

6.04 The Qualia Problem, the Cable Guy Paradox and the Conway-Kochen Theorem

Abstract
The Free Will Theorem of Conway and Kochen (2006) does not appear to fulfill its claims. The possibility of other mathematical and AI simulations, quantifications and descriptions of free will, emotions, feelings, physiological drives and the like is also briefly discussed with a view to clarifying that fact.

Time Travel in Theory and Practice

Recently Conway and Kochen (2006) produced a paper called "The Free Will Theorem". In the event, they claim therein "If indeed there exist any experimenters with a modicum of free will, then elementary particles must have their own share of this valuable commodity."

Now this should be an important paper and its publication was widely heralded as for example in the "Daily Princetonian" in 2004 where Nobel Prizewinner Phillip Anderson said "I think it's a positive development. I always thought the study of hidden variables should go, anyhow." Strong stuff. And now we have the actual paper, as distinct from the newspaper blurb.

I expect the paper will be widely read and could be used for many things, including attempts to prove that people do not have free will at all. This is even politically important as clearly it could in some way be used to excuse such practices as suicide bombing, a practice which Iranian and American psychologists (Pyszczynsk and Abdollahi 2006) now seem to claim is in some sense endemic in the psyches of both Americans and Iranians.

Now I will look at the Conway-Kochen Free Will theorem in a mildly roundabout way. I will start by considering something that looks a bit different, namely the Cable Guy Paradox (Hajek 2004). Some people could say that the Cable Guy paradox is only a pseudo-paradox but anyway we will still consider it.

The Cable Guy Paradox

Before beginning my discussion of the Cable Guy paradox I have two important points to make. Firstly, that the history of paradoxes suggests that, over the centuries, feelings are often different as to whether some so-called paradox really is a paradox or is really important or not. A case in point is Zeno's Paradoxes which are undoubtedly regarded as important today but whose value, relevance and meaning have been considered of variable merit or relevance over the centuries. We can look as far back as Parminedes and the Pythagoreans or as far forward as Chaitin, Russell and Grunbaum on the subject of Zeno!

Secondly, specifically regarding the Cable Guy paradox, we are primarily concerning ourselves in this essay with the two persons who make the wager -the wagerers- and their feelings during the matter. My view is that, whether we should consider feelings or not during the posing of the paradox, that having set up the paradox then during the morning is the time where the wagerers can be expected to have one view and the afternoon another view. And if, like me, you are not a person who likes a gamble, in Hajek's version you will prefer afternoon (as Hajek does) and in Restall's version you will prefer morning (as Restall appears to do). I feel that it is perhaps an unwarranted deduction at this juncture to deduce that neither Hajek nor Restall are themselves very keen on gambling (I do not know either) but it is an interesting deduction to be able to consider making so easily from the nature of a paradox.

The cable is to be installed by the cable guy during the morning between 8am and 12noon or the afternoon between noon and 4pm. Before 8am, we bet on whether he'll come in the morning or the afternoon. Hajek suggests that you should perhaps choose the afternoon as - very roughly summarising - it can be conceded that during the morning you will have less time left (perhaps an infinitesimal amount less if the cable guy is very early) which can thus make you regret a morning choice. This follows Hajek's own 'Avoid Certain Frustration Principle', stated here as given but later revised in the same paper (Hajek 2004): "Suppose you now have a choice between two options. You should not choose one of these options if you are certain that a rational future self of yours will prefer that you had chosen the other one – unless both your options have this property." However Greg Restall (2006) points out that "I don't know that you won't mug me, put me in the cupboard and mind the house without my company. Furthermore, you might record the front door with your camera, and then, after 4pm, let me out, and play the recording backwards. In that case, there's some interval of time in which I'll prefer having chosen the morning rather than the afternoon, by completely symmetric reasoning."

From our present standpoint what seems to be important is that we have actually simulated an emotion (call it regret, perhaps) and related it directly to a probability. In the Hajek version of the paradox, regret will alter the weighting of probability of choosing morning rather than afternoon, which on the basis of the proposed experimental setup would seem to be equal from a practical point of view, leaving aside the idea that the proposed time interval was an open interval in the morning and a closed one in the afternoon. So we have a physical measure of regret and an apparent paradox that we should wager on afternoon, despite the idea that the initial mathematical odds look the same for morning and afternoon. In the Restall version of the paradox, afternoon could clearly be a better option. We could deem the Restall situation unlikely but since we do not have specific figures as to how we should weight regret into the wager - the form of any expression might be complicated - we are simply left so far with an uncertainty effect due to possible (video) time reversal.

But odds which apparently are mathematically identical are changed on the basis of emotional views in a way which should be quantitative.

How can we place an actual measure on regret in such thought experiments ?

Sanders (1989), like many others, has tried to contrive a method. Sanders in fact focuses on a cluster of actions to do with ownership or possession of property but is able at any rate to show that the logic so expressed is able to handle many sets of circumstances. So in principle we are likely to be able to give a weight to emotion, and thus to calculate some mathematical results for the Cable Guy paradox which may differ from equal mathematical odds for morning and afternoon. So we undoubtedly can have, in principle, a mathematical way to handle regret. Of course, however sophisticated and accurate the methods we use for this become, we know that we are still just doing mathematics and logic. We are dealing with the map, not the country, as it were. In the real country, as I point out below, we seem to need quales, at least. As Goguen (2005) points out importantly and in more detail , "In Western culture, mathematical formalisms are often given a status beyond what they deserve." And even you wanted to give such formalisms a high status, you would have to give good reasons which generally do not seem to be available at this time. Goguen (2005) suggests that there are a number of objections to willy nilly equating the map to the country in such cases. In my opinion not the least of these is Harnad's symbol grounding problem.

Now we may also envisage the scenario where the Cable Guy and the person waiting for cable installation are respectively replaced by a laboratory experiment and a mouse - or indeed the mouse could even be replaced by one of Dennett's bats. In that example if we wanted to, we could even take the view that Dennett's approach rather than the use of qualia is correct, and we would not lose our argument by using some modern form of multiple drafts and not qualia for the reason that when we speak of free will for animals in cases like Dennett's case we may well be speaking of anthropomorphisms but we are still speaking of real possible scenarios or parables. However this is one case where the qualia approach looks simpler to me so I stick to that. It is reasonable to say that we can imagine a horrified mouse, whether anthropomorphically horrified or not. It is completely unreasonable to consider N horrified particles with N a very small number (possibly unity), anthropomorphically or not as the idea of N horrified particles would fall well outside the common or normally scientifically accepted view or parable (as Turner (1996) might call it) and totally outside the concept of freewill.

What has this got to do with the Conway-Kochen Theorem ?

Specifically, Kochen and Conway claim to show that particles' responses to a certain type of experiment are not determined by the entire previous history of that part of the universe accessible to them. But, even given Kochen and Conway's assumptions which I leave for others to debate at this time, all Kochen and Conway then seem to do is to allow an element of random behaviour to the response of the particles, a common enough assumption in mathematical physics, which has nothing to do with giving the particles free will ! Free will simply does not fit into the parable. Conway and Kochen do not appear to be singing from the same hymn sheet as everyone else.

We already know from the Cable Guy paradox that apparently identical mathematical odds can lead to differing appropriate odds in a wager due to emotional effects and we even know roughly (we think) how to calculate mathematically such odds. But given for example the view of Ramachandran (2003) "Obviously self and qualia are two sides of the same coin. You can't have free-floating sensations or qualia with no-one to experience it and you can't have a self completely devoid of sensory experiences, memories or emotions", obviously these emotional effects involve personality and qualia or whatever theoretical approach, probably even Dennett's approach, you want to replace them with. What these 'selves' most certainly do not appear to be, is simply small systems of partially random moving particles and little or nothing more. Such systems cannot be dignified as relating to or substituting for, quales. Quales are still a bit of a mystery and Conway and Kochen's particles do not describe them in any way whatsoever. If we did not want to tackle the matter from the qualia angle it seems likely that similar insurmountable difficulties would be encountered however we go about things. Simply, Kochen and Conway have not established the equivalent of free will for particles.

Indeed if we wish we can simply take their position as being an argumentum ad ignorantiam fallacy, that is that they assume that since something has not been proven false, it is therefore true. The classic case is that "Since you cannot prove that ghosts do not exist, they must exist". They are left with a few snippets like Conway's 'Game of Life', which is much less universal and broad than their present cases, and even the 'Game of Life' has no true

190

anthropomorphic qualities that anyone can detect. But they are the ones who have to prove their point at this juncture, not for others to defend themselves against their extremely vague hypotheses. Certainly if they could give a serious proof or substantial observation, most people would be happy to consider it, but the problem is from their standpoint that there is no reasonable suggestion that apparently randomly moving particles do imply free will. And they do not prove or even explain how or that such particles do imply free will. There is nothing to build on in what they actually say.

Donald (2006), who as it happens espouses a many minds theory, seems to have summed the matter up correctly yet again by saying: "They deny that what they mean by free will is simply neural randomness. They assume instead the existence of an 'active kind of free will that can actually affect the future' ", but they do not explain what sort of extra-physical homunculus can make active choices without input from a physical past." The 'extra-physical homunculus' which Conway and Kochen now seem to have tried to leave us with as our unwanted and un-needed fardel can, we might well hope, be dispensed with eventually - if indeed we need to consider it at all.. For example, it might be dispensed with by using category theory as I referred to in previous blogs or indeed by using many minds theory. But I think that the philosophical ideas posed by Kochen and Conway (2006) specifically on free will may not merit consideration as such.

So without a method like Donald's many minds, or indeed my Category Theory approach or even some other totally different approach, we can be still left (without good philosophical reason) by Conway (2006) with a lot of problems surrounding free will and none of them whatsoever actually resolved by Conway (2006). Even with either Donald's (2006) approach or my approach, we are certainly still left with a lot of things to do about free will but Conway (2006) has done none of them.

References

Conway J., and Kochen S., (2006) "The Free Will Theorem" , arXiv: quant-ph/ 0604079 v1, 11 Apr 2006

Daily Princetonian, (2004) Wednesday, November 24, 2004
http://www.dailyprincetonian.com/archives/2004/11/24/news/11569.shtml?

type=printable

Donald M.J., (2006) http://www.bss.phy.cam.ac.uk/~mjd1014/readings.html

Hajek A., (2004) "The Cable Guy Paradox", Analysis, Vol. 65, No. 4, April 2005 ;
http://www.nottingham.ac.uk/journals/analysis/preprints/HAJEK.pdf

Goguen J., (2005) www-cse.ucsd.edu/~goguen/pps/fois04.pdf

Pyszczynsk T., Abdollahi A, et al (2006), "Personality and Social Psychology Bulletin" 2006 32: 525-537 and news.yahoo.com/s/space/20060424/sc_space/thoughtsofmortalityturnpacifistsintokillers

Ramachandran V., (2003) Reith Lecture(5)
http://www.bbc.co.uk/radio4/reith2003/lecture5.shtml

Restall G., (2006)
http://consequently.org/news/2006/04/20/on_the_cable_guy_paradox/

Sanders K.E., (1989) "A logic for emotions: a basis for reasoning about commonsense psychological knowledge", Technical Report 89-23, Brown University Computer Science Department, 1989. (an expanded version of the paper: A logic for emotions: a basis for reasoning about commonsense psychological knowledge. Proceedings of the Eleventh Annual Conference of the Cognitive Science Society , Ann Arbor, Michigan, August, 1989)

Turner M., "The Literary Mind", (1996), Oxford University

6.05 New quantum approach to qualia, consciousness and the brain.

Abstract: In this paper I do not rule out the possibility of including the consideration of results such as those of Jahn, Walach, Radin and others, in

the spirit that I feel that, unlike much early USA scientific and technical opinion, I would not have effectively ruled out the possibility of the Wright Brothers as having discovered aviation. At the same time such results are certainly not paramount in considerations at this time. My approach uses category theory and a McTaggart A series as well as the conventional B series effectively used by Deutsch, Bohm and Penrose. This sounds philosophically and physically more realistic but at the present state of the art it may be required that the A series is a proper class. My theory will relatively easily link with any physically meaningful and duplicable NDE results which may be provided for example, by NDE experiments like those of Fenwick and Greyson and has many other advantages. Dream precognition results and ESP are very much denied by skeptics and on the whole by physicists. On dreams I certainly have not obtained precognition as such but noted apparent peculiar effects not dissimilar in superficial appearance. In psychology it is necessary to remember that many conclusions have been drawn and are repeatable from work like that of Strogatz. I favour dynamical systems psychology somewhat along the lines of Lange, but requiring an A series philosophy. By adding some ideas due to Stickgold and Hobson, I have already obtained preliminary surprising results. Presently I am proceeding to look at a structure somewhat along the lines of the Sprott work on psychology. I believe that through ignoring the McTaggart A series or trying to subsume the A series to within the B series, important opportunities are being lost and that early calls on quantum theory may be being made, when complex system theory could be more directly appropriate. http://ttjohn.blogspot.com/ presents the entire blog to date, including more work than that required here. The simplest appreciation of the situation may be that the present approach contains a past, a present and a future without further ad hoc additions and so in a sense exhibits qualities, generally recognised as certainly existing in human consciousness, but which are not obvious in theories which do not. Also it allows the existence of a God or Gods and free will (or indeed hypothetical gods or freewill) within its bounds, though does not insist on their existence a priori and in this sense it is more appropriate to consciousness theory than a conventional physics theory would be which almost excludes these factors or a theological theory which a priori insists on them. The absence of the possibility of freewill in a physical theory suggests solipsism or incompleteness rather than some disproof of free will and this is carefully avoided with the present approach, which yet contains much mathematics including all of quantum theory and high energy physics, together with chaos and catastrophe theories where relevant.

6.06 Self-Organised Criticality - a possible tool for the MBI

(1) Some neurological notes on our method of approach

First some brief details of current results in cognitive dissonance.

Children and monkeys both show 'cognitive dissonance', which has been - and still is in some theories - perceived as being very complex.

Given alternatives A and B and both of equal value (details in Egan's paper) they were asked to make a choice on the basis that they were not of equal value. Then they tended to derogate unchosen alternatives, when faced with a selection of a third item C in place of either the chosen or the unchosen alternative.

There is some white noise in the choice but overall the results were reasonably good.

When the choice is between A and C or B (not-A) and C, a simple computer could give the expected result for the derogated alternative.

Complexity of neural pathways in adults allows more mentally complex thoughts, perhaps, which simply overlay the above simple idea.

So in that sense, relatively simple 'computer-like' choices can be made for a brain model - in fact we had previously used 'Berkeley Madonna', and some results are online at http://ttjohn.blogspot.com/ (RSS feed available) and on a compact disc.

But if, for example, in the simple choice alternatives above, we had used instead a multiple procedure of choice with sophisticated undergraduates, the experiment would have led to some very different statistics, as in Eisenstein's example. There, there are clearly sophisticated reasons not found in the basic example.

A particular other case would be sarcasm - apparently a highly sophisticated mental process. According to the results of Shamay-Tsoory we should be able to put this into a suitably revised version of our Madonna program.

i.e. To interpret a statement literally, the right hemisphere is used (it is averred), the frontal lobes process the emotional content whilst the prefrontal cortex interacts with both. This was found out partly because posterior brain damage seemed to have little effect, and from other such clues. Whilst there are only 25 full studies reported, there would be further anecdotal evidence. All this seems to fit in with a basically simple 'Madonna' model still being satisfactory.

But we can go further and consider episodic memory. Rosenbaum's work seems to indicate that theory of mind (ToM) can work independently of episodic memory. Results come from participants with strongly impaired memory. There were so far only two good subjects but Tulving is a co- author of the paper and even one result would almost do. And also effects like sarcasm were still there in the chosen cases. It amounts to enough empathy to put oneself in someone else's shoes.

In fact Rosenbaum even said that "If you try mentally to put yourself in someone else's shoes, you do not even need to put your own shoes on, mentally speaking".

So a very simple theory involving a brain working like a computer with no long term memory provides a model.

Now we can use a simple 'Madonna' model to describe the brain mathematically in one of our 'bubbles', in the so-called 'pseudo A series'.

We do not need even to consider to begin with the work of Honderich, Libet, Klein, Trevena, Haggard and the like. And of course we have dealt with McTaggart's paradox in the MBI.

And it is hoped that we may be able to reach a basic philosophy of approach as simple as as, say, Winfree's, at least to start with.

(2) Some philosophical notes on our method of approach

According to Itano (2006), Zeno of Elea (490-430 BC) said that 'an object can only be in one place at one time. It cannot be in two places at one time. But to be in one place at one time is to be at rest. Therefore, the object is at rest. Therefore, motion is impossible'.

Strangely, Zeno's paradox has echoes of McTaggart's paradox in this framework, and the super- Zeno effect (Dhar, 2005) seems to mirror this idea.

I have tried to roughly indicate in previous blogs how the results of both Kwiat and Dhar are perfectly lacking in mystery and philosophical paradox in the MBI approach. Indeed these could lead in time to the building of a successful quantum computer, and help in many mathematical problems (e.g. Wigderson, 2006).

We also have a case already laid out in the MBI to look at some of the viewpoints of Hansen (2004) who points out that current physics special relativity has philosophical problems with regards to both Zeno and McTaggart. We point out a general way in Note 1 how such problems evaporate, as does the mystification with them, in the MBI.

At this point perhaps we should ask. Exactly what are we doing ? Well Frigg (2002) perhaps gives some pointers as to that. Frigg takes the view, for example, that the sandpile model, as such, does not fully describe the space-time view. He says, quoting references and reasons, "I don't think we are entitled to say that real sand-piles do exhibit Self-Organised Criticality". Well, whatever view is taken on that statement we are certainly only making sketches which can misdescribe at their extremes and give only a (useful) impression of reality. This is perfectly in accord with the thinking of the MBI. Frigg in fact says "often when a model is presented, only the briefest suggestive remarks are made about its bearing on the world ... one has to start somewhere, and quite often a false model may provide a good point of departure. The strategy then is to ameliorate the false model and to build a better one on its basis. Iterating this procedure leads to a series of models in which every one is an improvement of its predecessor. In this way, a simple and false model can initiate a series of models of ever increasing complexity and accuracy..... a model, even though clearly false, may lead us to think differently about certain problems, motivate new questions, shed a different light on some issues, and finally make it easier to adopt an altogether different point of view. In doing so, the model acts as an antipode to stagnant assumptions, undercuts too readily accepted hypotheses, and helps to 'de-familiarise' deeply entrenched styles of reasoning. In short, a false model can indicate alternative ways to deal with a phenomenon".

In other words, in the MBI we are acting more precisely by effectively specifying that we have a "pseudo A-series", which from earlier notes we consider we can use up to a point to mirror the A series using effectively B series mathematics. And in knowing what we are doing, we are more likely to be able to put matters right if and when the proposed mathematics goes astray with respect to physics as it really describes the world, or as we try to make the physics do.

(3) Some further possible work (potentially in progress)

(a) We can now consider further simple ideas like using the sandpile analogy as it has been tried, for example, with software development, without the physics actually disappearing from the system as the actual software used for development does in the paper (Wu, 2002). That contained an excellent analogy:

Driving force / sand drop /change request
Response / sand slide / change propagation
System state / gradient profile / release, iteration plan
Relaxing force / gravity / stakeholder satisfaction

but plainly "the mathematics was not the territory" just as "the map is not the territory" to a geologist. Even if a geologist goes along with the mathematical fractals approach, it does not get the dirt under his fingertips. But with the MBI approach in physics, we seem to be as close to a physical simulation of the real world as we can be at the moment. Importantly, for example, we have not simply beaten off McTaggart's paradox but on the contrary, we we have used it as strongly as we can.

This may well give us the ability to prepare a more precise or even a new and better Madonna model than the model N003b suggested in an earlier entry, using for instance some of the methods of Dhar (2006) particularly as described in Dhar's section 3 onwards and other such SOC methods, as well as what we are using to date.

(b) We know that Stickgold discovered a postcognitive effect by studying dreams, and used it in a very simple way, by stimulating the subject through the game of Tetris, and thus in turn controlling his future dreams. Our studies indicated that there is a retrospective effect as well, but existing models like N003b suggest that the effect will need to be finely tuned, with due consideration of the model used. At the moment a further simple test of individuals is planned, but the so-called precognitive results may or may not arise quickly. In the light of some field studies I made with some of Jeffrey

Gray's allegedly mildly synaesthesic patients, of informal discussions with Colin Blakemore and Simon Baron-Cohen, and other matters, the conclusion was reached that right now mildly anomalous but indicative effects like popout synaesthesia are often hard to achieve and detect. Indeed Blakemore (2007) indicated that he only knew of two popout synaesthesic cases in the entire world, one of which was also known to me, and both of these cases are being studied in America. That would confirm my results on the matter of synaesthesia so far.

The present work does not seem as if it needs special subjects, in fact we are more or less using any generally suitable subject at this time but it has become clear that a carefully planned model may be needed to enhance the reverse Stickgold effect. Results for both the forward and reverse Stickgold effect should be of interest in neurophysiology and both come under our remit as scholars. The fate of the MBI does not depend on the success in discovery of a reverse Stickgold as the MBI proves itself already as outlined above, but it would be a unique prediction for the model, confirming yet again its relevance.

(4) Easier observation and/or enhancement of Stickgold effect and reverse Stickgold effect

From a purely mathematical point of view, one could perhaps include in a more detailed way in N300b the fact that there could be tiny pushes and impulses to the mind at a given time, from both past and future stimulations, but that at a particular time the mind is in some kind of dynamic balance which Stickgold has altered in the 'Tetris dream' by a push from the past, relatively easy in retrospect. In my case I alter the position of the push from the past to the future, and this worked too. There is also plenty of anecdotal evidence for so-called 'prophetic dreams' but I have had discussions with Susan Blackmore and we are both of the same mind that this is not followed up with adequate statistics. But here we have a mathematical model which can be improved, applied to pushes from the future, and even checked by known pushes from the past which were confirmed to work at Harvard by good and acceptable research. We got the future pushes from my model which is still in a period of development.

Some kind of sandpile model might give further success. Obviously my preliminary Madonna model can be much improved. The earlier work is already on my website at http://ttjohn.blogspot.com/ and a CD and notes are available.

Simple attempted enhancement of these MBI models like N003b must be possible also by, for example, playing Tetris on a series of nights and trying longer and shorter times of play and times of day. There are other forms of stimulation which can be used. 'Alpine Racer', a skiing game of the early 'virtual reality' type, seems to have worked even better for Stickgold, perhaps because of the controlled switchback effect of such sports games and other factors. It is clearly important what effect on the visual field particular modes of presentation can have, but other factors such as the time in advance or after of the stimulus to the dream must be important. Also there is the level of dream recall by the subject, and the time of the dream during the sleep.

References

Blakemore C., (2007), personal communication.

Dhar D, (2006), "Exactly Solved Models of Self-Organized Criticality", http://theory.tifr.res.in/~ddhar/leuven.pdf

Dhar D., L. K., Grover L.K., Roy S.M., (2005), "Preserving Quantum States : A Super-Zeno Effect", http://arxiv.org/pdf/quant-ph/0504070

Frigg R., (2002), "Self-Organised Criticality - What It Is and What It Isn't", Centre for Philosophy of Natural and Social Science Measurement in Physics and Economics", Technical Report 19/02

Hansen N.V. (2004), Organization, "Where Do Spacing and Timing Happen? Two Movements in the Loss of Cosmological Innocence" 11: 759-772 ; "Spacetime and becoming", http://www.nielsviggo.net/philwork/

Itano W.M., (2006), Proceedings of the Sudarshan Symposium, Univ. of Texas, November 2006 ; arXiv:quant-ph/0612187v1

Wu J., Holt R., (2002) "Seeking Empirical Evidence for Self-Organized Criticality in Open Source Software Evolution", Available Research Reports, David R. Cheriton School of Computer Science, University of Waterloo, Ontario, Canada

Wigderson A., (2006), "P, NP and Mathematics –a computational complexity perspective", Proc. of the 2006 International Congress of Mathematicians. Madrid, August 2006

Note 1.

As pointed out earlier, without even invoking quantum theory, Quentin Smith (2002) explains how some models of the A series can seem to have B series concomitants, even in special relativity. In fact if we wish, we could consider our A series bubbles corresponding to different Tm values to be linked to one another by a spider web of gossamer chains. The spider web could now seem to be very clearly savouring of the B series, although we had started with a model based on the A series. Given suitable provisos, that spider web might well suit STR (Special Theory of Relativity). Now to quote Hansen:"One of the central theories of modern physics has particularly fascinating consequences when compared with concepts of time reflecting ordinary daily life experience: the Special Theory of Relativity (SR). The way SR is at odds with the classical idea of dynamic time or passage of time is not by implying that ideas of change or passage are themselves self-contradictory, as in the classical atemporalist arguments from Zeno to McTaggart. Rather, SR dissolves a necessary condition for the classical idea of passage: the existence of a unique order of the events of the universe, the allocation of every event to a point or interval on one axis of time over which the passage of the now might take place. (Or, with the equivalent inverted metaphoric of passage preferred by some, what is dissolved is the sequence of 3-dimensional "pictures" constituting a universal movie which might pass across the "projector" point of temporal presence.) In SR's reorganized grammar of spatiotemporal relations, events can no longer be said to be placed in such a 1-dimensional continuum of time and in a separate, independent 3-dimensional continuum of space, rather they are placed in a 4-dimensional continuum of spacetime allowing for a multiplicity of equally valid formulations of timelike and spacelike orderings, relative to velocities. This reorganization seems to complete what Bergson aptly phrased "the spatialization of time" so that the passing of the "now" becomes not only foreign and irrelevant in the physical universe but even cannot be formulated coherently in the context of physical theory. Apparently the result is a direct contradiction between our systematic knowledge of time as part of the structure of the physical universe and our intuitive notions of time, based on whatever unsystematic and perspective-dependent view of a fraction of the physical universe our immediate experience covers." Clearly this problem does not arise if we use the MBI.

7

Groundwork and Details

Chapter 7 covers the theoretical and mathematical aspects of the MBI not yet deal with in this book, and helps to spell out in a little more detail how time travel was developed here.

Section 7.01 is a few notes on emergence and abstraction.

Sections 7.02 comments in a preliminary way on McTaggart's A and B series and how they relate to modern science and neuroscience in the 21st Century

Section 7.03 points out how important descriptions of the mind using physics have seemingly not adequately represented ideas like freewill, and these have largely been effectively ignored or devolved in some sense to the 'folk psychology' and 'folk philosophy' realm. In this work we have begun an attempt to resolve these problems. We also discuss relations of the neural correlates of consciousness and work of Dylan-Haynes to my own MBI which has a mathematical model (outlined in Chapter 7) using the unconscious mind and complex system theory.

In Sections 7.04, 7.05 & 7.06 we try to deal with some of the complexities raised by the presence of the A series.

In Section 7.07 we attempt to find some phenomenal ways to deal with the above problems and look for sources of physical measurements.

In Section 7.08, the Many Bubble Interpretation and various examples of its use and effectiveness are referred to. The Schrodinger Cat paradox is essentially resolved in principle, the effect interpretable, Kwiat's recent result referred to, and the newly discovered described.

In Section 7.09, category theory, using some of the ideas of Varela, is advanced as a tool to study the conscious and the unconscious mind in a way which will allow the use of dynamic systems theory and the McTaggart A series. It is applied to the waking and the sleeping state.

In Sections 7.10 & 7.11 the dynamic systems methods as described in the books of Hannon and Ruth are used to do the actual physical modelling for the sleep and waking states.

In Sections 7.12 the more 'philosophical' - and mathematical - aspects of our theory are discussed with a view to still more future applications, and for use in our present efforts. Section 7.12 also relieves an Angst-ridden world of the problems of McTaggart, the details of which are also discussed on our blog and website.

7.01 Emergence and multiple-drafts

David Deutsch's "The Beginning of Infinity" is an interesting and thought provoking book.

Deutsch (1) points out that even in physics, reductionism simply is not fully applicable. He restates the interesting example of describing the position of a copper atom on the nose of a statue of Winston Churchill in London, England. in this case the theories of such ideas as war, leadership and tradition apply in a fairly direct manner but low level physics and the like do not seem to immediately assist explanation at all. He then goes on to point out that even in the case of describing in the case of the cooking of an egg, low level atomic physics is not much use for reductionist explanatory terms. He then says (2), using Hofstadter's example of the relationship between the concept of prime numbers and physical computing results, that 'abstraction' does help to explain certain matters of the mind and brain.

Time Travel in Theory and Practice

In my view we certainly do not have to go on beyond this to assume that either abstraction or emergence (or both together) is some kind of panacea to such philosophical problems. The understanding of the existence of parochialism as being an apparently inevitable concomitant of abstraction, and the subsequent intended advantage of cleaning up the parochialism by the abstraction, does not in any way ensure that emergence is a relatively simple route to success. Such an easy assumption might well suggest a simple answer to the 'digital or analog' problem (4) which homespun philosophers frequently suggest. In fact the implication which can be taken from Deutsch's idea that the idea of error-correction (4) in some sense allows 'digital' to be an advance beyond 'analog' will certainly not satisfy some other physical scientists (3) and so per se tends to imply that presenting the idea of parochialism in such cases, though certainly worth considering, is a dubious argumentum ad hominem which it inevitably can easily become - and often does !

The concepts of digital and analog were invented to describe idealized models of human-designed machines, and are far too narrow to encompass the subtleties of living creatures, for example.

A particular relevant case of the above would be the idea of 'freewill' being termed 'homespun', simply bearing in mind the now widely presented idea that the human mind can decide what it is going to do as much as 10 seconds before it becomes aware of this (5). By now such a denial of freewill is a guess belonging to a larger system of tacit assumptions about 'how the mind works'. I suppose that such a denial of freewill could be also construed as 'shoddy reductionism'. In short those who accept that 'freewill' is 'homespun' could themselves be described as 'homespun philosophers trying to take the over easy route'. That really does lead to a fairly gross acceptance of ideas like 'multiple-drafts' which really cannot help the case for multiple-draft enthusiasts when the matter is properly pursued.

So asserting 'homespun parochialism' is not enough, and certainly cannot allow an easy acceptance of emergence as a cure-all. Of course emergence is nonetheless a powerful and useful tool, just like pure mathematics is.

Emergence, abstraction and reductionism are tools which work, up to a point, but all have limitations. And certainly we must not throw aside ideas like 'multiple-drafts' but at the same time they certainly cannot be accepted as a sine qua non or even as a useful working hypothesis in some cases. Indeed we have to be careful not to use emergence as a philosopher's bootstrap.

References

1.Deutsch D., (2011), "The Beginning of Infinity", p109, Allen Lane, ISBN 978-0-713-99274-8

2. ibid., p 115

3. Andrade E., (2006), "The Organization of Nature: Semiotic Agents as Intermediaries between Digital and Analog Informational Spaces", http://www.library.utoronto.ca/see/SEED/Vol2-1/Andrade/Andrade.htm ; and Andrade's idea of 'form' is only one of many approaches to the digital/analog problem.

4. ibid 1, at 141

5. Soon C.S., Brass M., Heinze H.,Haynes J.D., (2008), "Unconscious determinants of free decisions in the human brain", Nature Neuroscience 11, 543 - 545

7.02 McTaggart's A and B series and how they relate to modern science and neuroscience in the 21st Century.

Abstract

McTaggart's ideas on the unreality of time as expressed in "The Nature of Existence" have retained great interest for 100 years to scholars, academics and other philosophers. In this essay, there is a brief discussion which mentions some of the high points of this philosophical interest, and goes on to apply his ideas to modern physics and neuroscience. It does not discuss McTaggart's C and D series, but does emphasise how the use of derived versions of both his A and B series can be of great virtue in discussing both the abstract physics of time, and the present and future importance of McTaggart's ideas to the subject of time. Indeed an experiment using human volunteers and dynamic systems modelling which was carried out is described, which illustrates this fact.

In this brief essay I will leave aside the notion that McTaggart's work has been exposed as being wrong, as in Maudlin (2002). Gale's work on "The Philosophy of Time" appeared nearly 40 years ago and much of this work was about McTaggart's approach. In the early summing up of Gale (1968) for example, and often enough elsewhere subsequently, the choice of A series or choice of B series or choice of either A or B series as desired is suggested. The great Arthur Prior, for example, as I remember him, had a clear penchant for the A series - and when I worked with Robin Gandy I remember that he found the A series at least acceptable.

Later workers, particularly in mathematical physics, tend to gravitate to the B series without really paying much attention to abstruse details. Simple special relativity or even simple dynamics will often solve the problems of such workers to date and these kinds of models almost give the impression of begging for the B series - or even simply ignoring McTaggart as a whimsical philosopher - quite wrong in my view.

When general relativity or loop quantum gravity rear their ugly heads - variations of simple 3D or 4D (space time) descriptions do not easily let most people pause for an A series. Philosophers can see value in the A/B series distinction in principle but often prefer a form of B series (Oaklander's (2004) "new theory" for example), (Note 1). This approach neatly fits in with conventional physics, but also with the idea that such apparent deliberate blinding to reality is perhaps a neo-Tychonian heresy, Margolis (2002). The problems tend to come out perhaps in psychological rather than clearly philosophical areas and I am thinking particularly of such things as Wason selection and even the work of Tooby and Cosmides. I have looked at these areas (Note 2) These matters are worthy of study in more detail and I will come to a few points later.

Briefly what I am saying, and obviously there are many provisos, hedges and restatements, is that I consider a form of A series and a form of B series - but the A series is probably a proper class (like the class of all automata for example is a proper class (Adamek et al, 2004) and probably cannot be effectively mapped, certainly not one to one, onto the B series. We only need to look at say Goldblatt (1984). In other words time is a rather complicated entity and when we get down to the mathematics or even the logic of time, we are using at least two different and not one to one mappable onto each other (mathematical) categories, A and B say (Note 4). And McTaggart had thus considered that he had found unreality in time when he was actually trying to compare two different things - though it is possibly not necessary to follow through his precise reasoning here, particularly on the C and D series. Clearly when considering a complex entity like 'time', there may ultimately be many more matters to consider but that does not remove the fact that McTaggart had found at least two such entities, the A and the B series.

Again roughly, much of current physics can be described by the B series but human nature - where we know we each have a past, a present, and a future - can perhaps best be described by the A series. And Yates (2007) has used dynamic systems theory to get working models and to try to obtain specific results for them. The details are a little more devious but I have presented posters at the Salzburg and Budapest conferences on the mind and consciousness this year, and am giving lectures in Mumbai later this year. Some details are in the following URL (Yates, 2007), details start at top of page)

So I got to a point that a reasonable physical assumption seems to be that the A series is a Proper Class. Bays (2001), in "Reflections on Skolem's Paradox " says "if we start with a proper class which "satisfies" some finite collection of sentences, then the Skolem hull construction lets us find a countable set which satisfies the same collection of sentences". In other words, I can use conventional complexity theory mathematics to study the A series as long as I remember that I am no longer in "block time" or B Series time. Now authors like John M. Gottman (2002) have used simple enough mathematics for years to solve psychological problems and seemingly eventually tried to shoehorn their ideas rather without thinking into a "block time" scenario where they will not properly work. And the heartbreaking discussions amongst proponents of tensed and non-tensed time may never again have to carry so much weight, (at Sydney conference, (2006) they mainly did carry weight) in a situation where a tensed time (A series) is used where convenient for people, and a non-tensed time (B series) where convenient for objects. And the maths can be great in both cases. Though of course that does not say it will be straightforward or easy.

In a sense we can argue that people have been misled by the B series followers - the 'soul-less' physical scientists - whose methods undoubtedly work excellently in simple dynamic or relativistic situations - into thinking that the B series really is all there is to be used, and that results have to be niced down to fit in with B series dogma. In this regard one could almost place such people as neo-Lysenkoists - with equal certainty of failure in the long term, though minor agreement with facts if the mathematical models are not too bad. One is not speaking polemically, merely stressing the fact that a modern version of the Tychonian heresy brings many problems.

Time Travel in Theory and Practice

So now comes the crunch: Where are there, in the known physical or biophysical world, clear distinctions between A series and B series results but that A series triumphs ? To narrow down the field, first we may ask why we need the A series anyway if we are doing physics. One simple answer is that when physics added wave/particle duality to its toolbox, it became apparent that it did not describe the world as we know it. We know that people would take an entity to be either a wave or a particle, not both. Simple tricks like the Copenhagen Interpretation cover up the cracks. Certainly, but the cracks are still there. More complex ideas like many-worlds-interpretation may eventually lead to the discovery of alternative universes, but right now these haven't been found and the new worlds of science are already not the worlds of human beings. There is worse to come. And that is chaos theory and non-linear equations. After the work of May, Lorenz and many others, we know now that even classical predictability is uncertain. As they say, the change in the direction of a butterfly in Santiago today can change the weather in Toronto tomorrow. And there is not a way that we can, or indeed probably want to, change such minutiae. There is, though, a gleam of hope! We can make predictions about the duckrabbit in Note 2, for example. The duckrabbit problem has been solved to the point of quantitative assessment! The details are in Sprott & Aks (2003) and it has been done for the Necker cube. So these 'psychological' techniques are going to work quantitatively.

An objection could be that we don't care, that we see lots of optical illusions. What we want is a 'real' phenomenon that can be really observed. I have considered many possible phenomena of this kind. A good example, though, is hypnosis. Apparent visual and other mental phenomena which individuals claim to perceive under hypnosis are legion. These are sometimes bizarre and well outside the laws of physics, yet they are there - as Galileo said in his alleged historical quip "it (the phenomena) still moves (still is observed) ". Normally other people who are not hypnotised do not see them, but often enough, appearances to those who do can go against the present laws of physics for a long time. Now for years many people didn't believe hypnosis is anything other than a simple artifact of the mind. But recently Kosslyn (2000) showed, using PET brain scans, that 'something' (physical) is happening in the mind when hypnosis is carried out, and that 'something' can, for example, change the appearance of a style picture to the hypnotised viewer. So we have a 'real' phenomenon that can be 'really' observed, by one or more real persons. Now to say that does not count, it is only in the mind, simply places the burden of doubt somewhere else. In terms of physics it is as 'real' as anything else. PET scans prove it.

We could have a grim choice of discounting observations coming from the area of the human skull as being 'unphysical', or alternatively placing them in a new class of mental-but-physical phenomena. It did not perhaps much matter if we assigned such observations as 'unphysical' when there were many fewer quantitative and accurate measurements of what goes on in the brain, but now, with fMRI and other new techniques, it does matter. It is also plain that the provenance of all sensations, real or imagined, should be regarded as suspect. The era of Dr Johnson kicking a stone with his foot as a way to disagree quickly and reasonably adequately with Bishop Berkeley is, for good or ill, now long gone.

There are clearly other many other such cases arising. In my blog I try to consider some of them. Near death experiences are perhaps one of the most controversial of these. Dreams are the most common example, perhaps, and for the moment it may be easier to manipulate or create these than to change other human sensations or to look for difficult cases, though needs must. For example Quinton (1973), when he criticises Kant, for example, specifically makes use of thought experiments on rather unusual and somewhat obsessive dreams. I must point out that whilst dreams are not the precise lychpin of Quinton's argumentation, at least they are an important part of the argumentation's thrust, so dreams definitely cannot be ignored as an arena for philosophical work of this nature.

Dreams are also of great interest in neuroscience as for example, it is now possible to preprogram people's dreams in advance, at least to a limited extent, as Stickgold (2005) has shown. A simple computer program using was therefore used to set up a model to describe this effect, and the results are shown on my blog at http://ttjohn.blogspot.com/ . Now the philosophical problems involved in predicting the future using a B series representation and such a model could be endless. A lot of counterexamples (Note 3) could readily be set up, in some cases at least reasonably convincing, in block time. But in the A series we have the luxury of not having to bother to do so, even though real restrictions will undoubtedly be likely to be apparent to its actual physical (not just philosophical) use in the long term. So for the A series we just need to consider a person at some phase, with a past, present and future which has no known necessary call to alter other bit of the A series, as yet unexplored, and to record dreams before and after a Stickgold perturbation. Then the same model with slightly different parameter values allows precognition within the model, which also occurred for our test case. In other words the unexpected dream 'prediction' of playing a game of Tetris occurred before the stimulation was made, by a subject unaware of the nature of the exercise.

Repeatability: I would consider the Milgram (1974) experiments to be perhaps the best experiments ever done in recent years in experimental psychology and they are highly predictable, over many experiments by many people. I am hoping for that level of predictability but so far we have only catalogued one test subject (of Indian origin) and will be looking for more suitable subjects in India during (2007/8). The results suggest that quite tiny parameter changes could greatly affect results and therefore these experiments may take a long time and if necessary co-ordinated with other related matters, and different experiments. What we do have, however, would seem to be a proof of concept in that ordinary modern mathematics can be used, and possibly ordinary experiments will lead to useful results. I claim only proof of concept so far. Out of interest, the present work is very recent but a patent on time travel was taken out by me on the topic, probably covering basic points, about 25 years ago, and I have let it lapse, so I trust that most serious future work may effectively remain open source, at least in the U.K.

References

Adamek J., Herrlich H., Strecker G.E. (2004), "The Joy of Cats", p15, John Wiley , and katmat.math.uni- bremen.de/acc/acc.pdf

Bays T., (2001), Ph.D. Dissertation. "Reflections on Skolem's Paradox" p86, UCLA Philosophy Department.

Gale R.M. (1968), "The Philosophy of Time", p65 et seq, Macmillan Goldblatt R, (2008), "Topoi", p10, Dover

Gottman J.M., et al., (2002), "The Mathematics of Marriage", MIT Press , ISBN : 0-262-57230-3

Goguen J. (2006), "Mathematical Models of Space and Time", http://www.cs.ucsd.edu/~goguen/pps/taspm.pdf

Kosslyn S. M. et al, (2000), "Hypnotic Visual Illusion Alters Color Processing in the Brain", Am J Psychiatry 157:8, August, 1279

LePoidevin R., (1991), "Change, Cause, and Contradiction", p34-37 inter alia, New York, St. Martins Press.

Margolis H, (2002). "It started with Copernicus", McGraw Hill

Maudlin. T. (2002), "Philosophers 'Imprint, Vol.2,No.4, "Thoroughly Modern McTaggart ". In that article. Maudlin says "Let's return McTaggart to his final resting place,and let him molder there in peace". Earman (up to a point) disagrees. And so it goes on. Works like the Stanford Encyclopedia of Philosophy still publish disagreements either way or any way. But the point has to be made that not all philosophers agree on the relevance of McTaggart's work, though that does not of course imply that counterfactual arguments will easily bear fruit.

Milgram, S. (1974) "Obedience to Authority", Harper & Row, USA

Mortensen, C., (2006),"Change", The Stanford Encyclopedia of Philosophy (Winter 2006 Edition), Edward N. Zalta (ed.), URL =

Oaklander N., (2004), "Freedom and the New Theory of Time" p337 - "From The Ontology of Time", (Amherst, NY: Prometheus Books, 2004)

Quinton, A, (1973), "The Nature of Things", p164, inter alia, Routledge & Kegan Paul, London.

Ryberg et al (2006), "The Repugnant Conclusion", The Stanford Encyclopedia of Philosophy (Spring 2006 Edition), Edward N. Zalta (ed.).

Stickgold R., (2005), "Sleep-dependent memory consolidation" , Nature, Vol 437, p1272

Swartz N., (2001) "Beyond Experience: Metaphysical Theories and Philosophical Constraints", Second Edition free at http://www.sfu.ca/philosophy/beyond_experience

Smith Q., (1993), "Language and Time", Oxford, Oxford University Press ; an important and additionally elucidatory discussion in Swartz (2001).

Sprott J. C., Aks D. J., (2003), "The Role of Depth and 1/f Dynamics in Perceiving Reversible Figures", "Nonlinear Dynamics, Psychology, and Life Sciences, Vol. 7, No. 2, April 2003

Stevens J. C. , (1984), Ethics, Oct;95(1):68-74

Time Travel in Theory and Practice

Sydney conference (2006), "Time and consciousness Conference", http://fragments.consc.net/djc/2006/07/time_and_consci.html

Tooley M., (1998), "Time, Tense, and Causation", Oxford, Clarendon Press.

Tooley M., (2006), http://spot.colorado.edu/~tooley/LectureforExercise4Phil3480.html

Yates J., (2007), http://ttjohn.blogspot.com/2007/07/2007-tsc-qmind-conferences-my-abstracts.html#links

Note 1

Two important recent books which illustrate this fact are Smith (1993) and Tooley (1998), both of whom seek to escape from the strictures of McTaggart, but in different ways. Very briefly, Smith uses a changing present which needs peculiar tensed (past/future) facts which change relative to their successive "presents". In short it is almost an A series. Tooley's accretion of times seems to require an infinity of instantaneous accretions of times for any finite change, or worse, a finite number of "timed" accretions of times. So Tooley, like Oaklander, is very nearly a B theorist. All these ideas are philosophically interesting but to me they don't trump Prior. LePoidevin (1991), for example, speaks of Prior's "temporal solipsism" but nonetheless points out that Prior's tense ontology is, in his view, perhaps the most plausible way to avoid McTaggart's contradictions while maintaining McTaggart's intuitions of the significance of tensed propositions. The newest relevant commentary may be Mortensen (2006).

Note 2

In my blog at http://ttjohn.blogspot.com/ in the article "Are Physicists suffering from the Tychonic Illusion ?" I place the incisive comment of Margolis "So are cognitive illusions likely to occur and be hard to correct beyond the narrow contexts of psychology experiments? I think that is inevitable, so that it is important that we come to understand these illusions. Careful thought about simple puzzles like Wason can help illuminate what happens in vastly more consequential realms." To put it crudely, it is almost as if the solar system were just a large duckrabbit unclearly unidentified or a Wason card test unsolved. [A duckrabbit is a picture which seems to mentally

212

alternate during viewing between looking like being a duck and a rabbit].

Note 3

The question of counterexamples and paradoxes is a major one for time travellers, and has filled many encyclopedia pages. We are left with problems with counterexamples of any type. Those commonly given by Tooley (2006) on other topics for example, have been disputed, often for possibly good reasons or sometimes apparently for unphilosophical ones, Stevens (1984). We are fortunate that counterexamples are not so prevalent so far in the A series. And indeed some would say that counterexamples go in our favour here, as even one example of time travel proves that it happens and we now have one example - which I mention in this essay and give details of in my blog. I would say that in principle there is a strong counterexample here, but in my view we do need more cases to pursue a convincing program. And I have come across many general problems with this "one example proves it" approach when considering the synaesthesia work of Jeffrey Gray with him, also discussed briefly at http://ttjohn.blogspot.com/ A further angle to the McTaggart work is of course ethics and the "repugnant conclusion" (Ryberg, 2006) as well as relating to the work of Parfit, which has to be omitted for the moment since we are largely pursuing the physical/mathematical side of the paradox in this brief essay.

Note 4

Of course category theory is at the crux of a proper mathematical explanation of the paradox, as Yates (2006) duly notes elsewhere. In a somewhat similar instance, the Buddhist Monk puzzle, perhaps first described by Koestler and mentioned by Fauconnier and Turner, a category theoretic explanation comes closer to the light of day in the mathematical discusion of Goguen (2006). Fauconnier and Turner use a network model which would need re-examination to apply to the present context. And indeed, for a spiritual interpretation, real Buddhist monks are at work right now but for present purposes we stick to mathematics, physics, neuroscience and philosophy for the moment.

7.03 Explorations of available philosophical ideas using modern observations

Abstract
For nearly a century, since the discovery of wave/particle duality, physics has not been a description of the world as we know it. With the event of QED and particle physics, even dogs seem to have more gut comprehension of physics than humans. Worse, for half a century, important descriptions of the mind using physics have seemingly not adequately represented ideas like freewill, and these have largely been effectively ignored or devolved in some sense to the 'folk psychology' and 'folk philosophy' realm. Current philosophical ideas must resolve this matter, and here we have begun an attempt to do so.

The work of Cleeremans (1999) and Haynes (2008) stress the importance of the so called NCCs or "Neural Correlates of Consciousness" in studying mental activity. Haynes (2008) provides physical results which some would say raise questions about free will.

In fact Cleeremans (1999) goes so far as to say specifically "philosophy of science may help and provide a metatheoretical framework for the current interdisciplinary project.... Indeed, the only assumption such an approach requires is that of a lawful covariance between cerebral and phenomenal processes"

This assumption in itself seems to presuppose a sound superstructure of theoretical physics, as the phenomenal processes are traditionally described in terms of current physics.

I remark in a current paper (Yates, 2009a) "free will philosophers either ignore Haynes's work, or deny free will already, or are seeking a work round. Fortunately I do not seem to need a work round as Haynes's work seems to provide simply more evidence that the McTaggart B series is insufficient and we need the A series as well." So, I am satisfied with Haynes's (2008) results and (generally speaking and contrary argumentation aside) with free will also.

In my view, philosophy can provide additional questioning which may be able to add further parameters to my mathematical dynamical systems model (which incorporates both the McTaggart A and B series) as well as the current results and so on whose value must be subsumed to philosophical considerations. This model is discussed in Yates (2008).

I believe my present model may ultimately solve many problems relevant to philosophy, in subjects like time and freewill. And I think it is already doing so. So the right philosophical queries to subjects and many other philosophical matters are of great importance to me.

I have been well aware of the work of Kornhuber, Libet etc., more or less since it was published, as founder and editor of the "International Journal of Theoretical Physics", for which I personally attracted many years ago the usual array of specialists and Nobel prizewinners. People like David Bohm, Roger Penrose, George Gamow and Louis de Broglie were on my editorial board. The journal is referred to occasionally in my websites, in particular http://ttjohn.blogspot.com/ . I needed to know of the Libet, Kornhuber etc. work for my fundamental studies.

Basically current physics is unfortunately completely quite inadequate. Dogs understand physics better than people (Orzel, 2009), and that gives no kudos to dogs but at a basic level may simply indicate that people are smarter than dogs and more reflective. Physics was accurate enough for purpose in the days of Newton and Einstein but today we live in a different world. For example, it is only weeks ago (Cubrovic, Zaanen, and Schalm) that the current very basic B-series string theory in physics may have been given a firmer foothold. A-series is mainly overlooked in practice.

Haynes's (2008) work obviously moves the work of Kornhuber, Libet, etc. forward another step. And to omit a proper consideration of the A series at this point is rather like trying to do timekeeping at relativistic speeds without special relativity theory. Timekeeping at speeds much slower than relativistic speeds clearly works well enough for its own purposes, but special relativity is obviously needed for the higher speeds. In the case of studies involving mental processes at the level of abstraction of say freewill or (if postulated) qualia, the A series, not just the B series, is needed.

I am hoping for some help in this regard. I am really trying to get some important new work done and I wonder how best to get this across to the philosophical community, and also to get more feedback for my own work.

Simple exposition of what we have done so far: The brain is treated not in a totally simplistic way as a wired up and complex computer or a bunch of neurons, but like a mind battling between objectives. For the moment 'conscious' mind is taken as 'Juliet' and 'unconscious' mind as 'Romeo'. Using Gottman's mathematical theory of marriage guidance counselling and attractor theory after considerations like those of Winfree and Strogatz, equations arise, as given in Yates (2008) and on the website. Further references are in Yates (2008, 2009) and on the website.

More complex brain models are of course possible and are welcome additions to any discussion. Primarily, just as marriage counselling requires actual discussions as well as measurements, the present approach requires experimental philosophy as well as fMRI readings.

Already we (Yates, 2008) have discovered the Reverse Stickgold effect, which seems to mean that we may dream about what we are going to do, as well as what we have done. In a way this sounds obvious, but the details are not so obvious as sometimes people seem to have no advance idea as to what they will do. More and more this may be coming into phase with current physics experiments such as the Haynes work (which tends to verify/extend Libet), and philosophy owes it to all not to allow scientists to throw out the ideas like free will without thinking it through, as they tend to be prone to do.

We had posters at three consciousness conferences recently, in Budapest, Salzburg and Tuscon but putting across useful work at such places is not easy. My website http://ttjohn.blogspot.com/ contains many of my current thoughts.

References

Cleeremans A., Haynes (1999) J-D., "Correlating Consciousness: A View from Empirical Science" , Revue Internationale de Philosophie 3 (209):387-420 http://srsc.ulb.ac.be/axcWWW/papers/pdf/98- NCC.pdf

Haynes J-D., (2008), http://medgadget.com/archives/2008/04/not_a_free_will_after_all.html

Orzel C., (2009), "How to teach Physics to your Dog", Simon & Schuster, ISBN-13: 9781416572282.

Yates J. (2008), "Category theory applied to a radically new but logically essential description of time and space", http://cogprints.org/6176/

Yates J., (2009), "A study of attempts at precognition, particularly in dreams, using some of the methods of ", http://philpapers.org/archive/YATASO.1.pdf

Yates J., (2009a), "Do Intuitions about Reference Really Vary across Cultures?", on my website http://ttjohn.blogspot.com/ at http://ttjohn.blogspot.com/2009/06/do-intuitions-about-reference- really.html

7.04 Memetics and the A Series (1)

When we look at the A Series and the work of, for example, Varela and Ehresmann as frequently referred to in earlier blogs, but consider problems such as (and only "such as", certainly not exclusively) that the A series may be a proper class (that is, roughly, a class which is not a set) we are left with at least two obvious approaches, the first being to examine further immediately the A series properties of time and consciousness in terms of the work of Varela, Brown and Ehresmann for example.

Now attempts (which unfortunately normally did not specifically invoke A series) have already been made in the literature to do something like this - I am thinking particularly of the work of Baas (1997, 2004), the work of Ronald Brown often referred to in this blog, and in particular a recent summary of the situation in the n-Category Cafe blog (Baez, 2006) and there is the Blog "Machine Learning Thoughts" (Bousquet, 2006). The latter brief blog entry makes it exceptionally clear that not only is a lot more work needed before machine learning can reasonably aspire to the rare and refined level of detailed philosophical debate, but we can see that the maths is not yet well enough founded even allowing for the fact that nowadays proofs or even establishment of ideas is in itself tied up in computerisation (I'm thinking Chaitin and further, also Note 1 in addition as other problematic areas), which of course could make it all far more complex.

So basically what simply looks like a relatively straightforward mathematical exercise (if undoubtedly an extremely difficult one) is truly fraught with complexities. In other words I am not just saying, OK it's a difficult topic but I am suggesting that perhaps the easiest option might be to take a different angle on the whole thing, even though in my opinion we have very much gone some way along the course by establishing the apparent need for the A series as well as the B series and suggesting that all this is mathematically and philosophically feasible. So we may have made some headway, which we will retain if possible.

What I am saying now is - lets look at the facts, find enough new scientifically establishable work, take a head on view that there may well be a real 'hard problem' as well as many other basic philosophical problems which we must at this time seriously confront, and keep it all very empirical to begin with.

Whilst nowadays many regard Carl Gustaf Jung as somewhat of a ladies' man and simply a verbose mystifier, he said a lot of smart things. In particular in his disagreements with Freud he brought out the idea of the Oedipus Stain - this meant that in a way Freud was considering his patients as if looking at a culture under a microscope and, by staining it, considering only certain features. Clearly if he had used a different 'stain', he'd have considered different mental faculties and of course up to a point Jung seemed to profess that he had a whole series of different 'stains' he could use or even, up to a point, no stain at all. So what we may need to do is to consider different mental perspectives on the problem of A series time, and see which if any of them produce the best results and even allow us to present an effective, even predictive, mathematical model. This could be either a true A series (which we would hope for) or a pseudo A series, perhaps even presented mathematically like a B series or concatenation of B serieses, as mentioned in earlier blog entries.

For this purpose at least, why not consider memetics?

Considering Memetics

I have been looking further at the idea of memes and genes, since it does look as if it could possibly fit in with the McTaggart A series in the way which I am using it at the moment, assuming that the A series is a proper class. Thus a complex systems explanation of the A series could help to tie in with experiment.

I note that memetics has a long way to go yet, for example Susan Blackmore's (1999) ideas in her important early work tends to run against problems with mirror neurons, which I think were not well known (Rizzolatti, 2004) at the time of writing, and certainly are now known to occur in animals and humans. There is other potentially hostile work such as some fairly recent work by Edmonds (2002, 2005), insofar as both his "three challenges" and the "revealed poverty" article are concerned but the "revealed poverty" article itself, for example, has problems as the "Journal of Artificial Societies and Social Simulation" which he refers to as to an extent providing more

218

successful replacements for memetics seems little more successful than the (currently discontinued but possibly to restart) "Journal of Memetics", and I know from many years of experience, high circulation of a journal does not relate to correctness, relevance or often even much to popularity and acceptability anyway.

So I think the meme idea still may have mileage and use, but the idea of memes being a sort of "Theory of Everything" may be a bit much as even physics is finding that idea difficult.

The early papers on Chaos Theory by Lord May for example, were pretty precise but maybe it is too much to expect that sort of precision at this early stage, I feel we are are still, relatively speaking, in the days of Lorenz.

Part 2 of this paper will provide preliminary results.

References

Baas N.A., A.C., Vanbremeersch J-P., (2004), "Hyperstructures and memory evolutive systems" International Journal of General Systems, Volume 33, Number 5, October, 2004, pp. 553-568(16)

Baas, N. A., Emmeche, (1997) "On Emergence and Explanation", Intellectica 1997/2, no.25, pp.67-83

Baez J., Corfield D., et al (2006), n-Category Cafe
http://golem.ph.utexas.edu/category/2006/12/back_from_nips_2006.html

Blackmore S., (1999), "The Meme Machine", p49, OUP, ISBN: 0198503652

Bousquet, O., (2006) Machine Learning Thoughts, "Making Machine Learning more Scientific",
http://ml.typepad.com/machine_learning_thoughts/2006/06/making_machine_.html

Edmonds, B. (2002). "Three Challenges for the Survival of Memetics".
Journal of Memetics - Evolutionary Models of Information Transmission, 6.

Edmonds, B. (2005). "The revealed poverty of the gene-meme analogy – why memetics per se has failed to produce substantive results". Journal of Memetics - Evolutionary Models of Information Transmission, 9.

Josephson B., (2001) Cavendish Physical Society talk 9 May 2001,
http://www.tcm.phy.cam.ac.uk/~bdj10

Kandel E.R., (2000) "The Molecular Biology of Memory Storage" Nobel
Lecture, http://nobelprize.org/nobel_prizes/medicine/laureates/2000/kandel-lecture.html

Rizzolatti G., Craighero L., The mirror-neuron system, Annual Review of
Neuroscience. 2004;27:169-92

Notes:

1. Brian Josephson, for example, has written papers on mathematics and brain
functioning, and without including a bibliography of such work, a brief
description as in (Josephson 2001) makes it clear that he has plumbed to the
depths of a lot of relevant ideas. Perhaps on a more practical note, workers
such as Kandel (2000) try to sort the brain out but my take on Kandel's work is
that in his own way he is a good deal more practical than say Penrose but he
certainly does not answer all our questions, or so far even point a finger as we
might want.

7.05 Memetics and the A Series (2)

We hope to establish the concept of a model or series of models which can be
developed mathematically and lead to consequences which, as well, can relate
to mathematically predictable facts, ideas, results or further enquiries.

We can look at points in the A series (or its derivations) which may be disjunct
mathematically but can have some relationships which can up to a point be
defined within themselves. Crudely, at each point in the A series or its
derivations, for each individual point we may want to write down a past, a
present, and a future,

Perhaps the obvious things to look at are the logistic map, (or some extension
thereof) and the which should lead to some temporal inequalities within the A
series. The classical case of the Hurst exponents is that developed by Hurst
himself and concerns flood levels on the river Nile. The historical structure led

to the construction of the Aswan high dam and ecological problems ensuing from it, these facts and the UK/French/Israeli Suez invasion 50 years ago. All these matters need to be considered if the historical context has to be properly explored, though we'll leave them aside for now.

We leave historical context aside for the moment and bear in mind the exposition in Peters (1991) (Note 4). Fig 9.7 in this book, for instance, shows how R/S and the Hurst exponent vary with time for the Standard and Poor's 500 index. It is important to remember that the R/S is time dependent and in theory (depending on how we scale the exponent) a value of 0.5 is a noisy value, between 0.5 and 1.0 means future dependence on past and between 0 and 0.5 means some kind of de-correlated future and past. Thus the Hurst coefficient in principle should allow us to set up a series of scalars - and we could start at one time and work to a future time or start at a distant future time and work towards an earlier time or a present time, merrily graphing the as we do so. In this way we could in principle reverse future and past.(Note 6 and beware Note 5)

Another point perhaps worth referring to at this juncture is that, as Schlather's (e.g. 2001) work clearly shows, there is no equation relating the Hurst exponent H to the fractal dimension D for all systems although some people seem to assume that the two things are necessarily simply related by $D = n + 1 - H$ (n being the normal dimension of the space). Hurst exponent is not the same as a kind of rescaled fractal dimension. (see Note 3)

Now Ian Posgate's Lloyds Syndicates, for example, give instances of contrived statistics which will have a different (or 'human') element to those of flood measurements on the River Nile. In one case nature gives us a set of statistics and in the other clearly human revisions will have given us the figures. The same is even true in something as simple as stock options, where the person holding the book (however bad a mathematician he may be) will have taken great pains to ensure that his commissions should compensate for any inadequacies in the system, making it plain that any mathematical computation in advance of say, stock options or casino odds is a losing venture, even if for no other reason than that large winners at casinos are usually banned unless for some such reason like the winners are simply shills. So money making ventures using Ruelle-Takens methods, for example, as Packard apparently did, are unlikely to be useful from our viewpoint and indeed are just part of the system, now in large part human, which we need to consider.

This does not mean that human enterprises cannot come under our scrutiny of course, quite the opposite as we are considering memetics not winning at casinos. In fact I take the view that the work of Ian Posgate and Nick Leeson come, as memes, directly and very appropriately under our aegis ! (This does not imply that we are looking for such things as economic prevention of such practices - or indeed the reverse - but rather that we see a meme qua meme). But it is possibly easier to begin by considering natural physical processes or events resulting from more humble living creatures than humans. This does not imply that humans are the only creatures capable of rational thought as some memeticists would like to claim, the question is still open at this point. Here we are just plain that a bird, a dog, or a slug seems to act more like a human than a drop of rain does, and that the statistical laws and other behaviour of such creatures will likely seem more 'human' than those of a drop of rain. This does not of course mean either that we are trying to outsmart non-human creatures, more that because their attitudes are different their attitudes are possibly less likely to conflict with our experiments. (Note 2).

To calculate and interpret them should be child's play with the aid of a great deal of free software (Note 1) already on the market. One might also expect that the human or animal might reveal a lot of information as to differences from simple 'physical' results (perhaps along the lines of Velman's philosophical postulates), a good meme leading to survival giving one result and an unsuccessful meme leading to another. One could also traverse the time scales from past to future or from future to past and obtain Hurst tables which by the way they are written could obtain different results forward in time' to those taken 'backwards in time'.

One clearly would expect living creatures to present a different showing of to non- living ones. Unfortunately the figures are far from clear, even for blowflies. Nicholson's pioneering work on the ongoing population of flies in a bottle, for example, is described in particularly clear lay terms in (1989) (p263, 270 et seq). Miramontes and Rohani (1998) have a newer take on this matter. ("The ubiquity of 1/f dynamics is one of the major puzzles in contemporary physical science" also see Note 7).

Rohani (2002) now suggests using more advanced methods than the Hurst exponent which, also in my experience, carries a lot of perplexities with it. It is also hard to get enough useful and interpretable data. This and other similar matters will be discussed in "Memetics and the A Series (3)" being written now, and which contains further approaches to the statistical problems which a great many people from Edmonds (2002) through to myself and Rohani see as important, such as the relatively large amounts of data required and the relatively meagre and uncertain results obtained.

Gatherer (2005)', and others such as Marsden and Lynch, suggests inclusion in a study of memetics items like terror and says it may help us to identify cases where our a priori thinking about a cultural phenomenon is inadequate. But he feels that the future of the real world can't be confirmed with a computer model. My own findings were, when I tried the more mundane statistics like London Tourism figures and the like, was that those tend to be tedious, similar and very often give the appearance of being massaged (though I reserve judgement on whether they are, or just look that way if one is seeking anomalies) there is certainly seems to be more to learn from the 'homebodies vs. the hellraisers' model which Gatherer seems to largely espouse. And I point out from my comments above on Leeson and Posgate that such matters should be important in the use of statistics. But in the cases of Leeson and Posgate it is unfortunately prime facie clear that statistics have been massaged and that is probably even true for some terrorist crime statistics. It is vaguely annoying when you think that you may be onto something good but find that quite a lengthy search for useful statistics in quite sane places seems to provide no useful results. ('We all knew that this figure or that one increased every year', someone may well say, and statistics correspond). I found US Bureau of Justice Statistics (2007) well worth considering though. They are worth running through the computer and quite easily I got Hurst values varying from about 0.51 for later homicide statistics (semi-random slaughter in statistical terms) to a more predictable 0.97 for earlier ones through 0.78 and .and 0.84. The interesting part of the murder figures is the dip between about 1940 and 1960 when the rise apparent earlier, appears again. On a cyclic basis maybe the rise should soon start again. If you reverse the statistics and start at the later end the Hurst figure tends to be very near unity. But here again there are only rough accounts possible.

These matters are proceeding - the A series is in rough lay terms something like a tensed theory of time as opposed to the (tenseless) block time of the B series which physics (up to a point) is so happy with. The idea of using scalar entities as various points in the A series representing (past- present-future) at different points in time is an approximation that we may have to live with for the moment. However to look into further steps to be considered, we can look at the work of Gabora ('strange attractors' and the like) on the one hand and Sprott (etc.) and Rohani on the other in our next installment "Memetics and the A Series (3)".

References

Edmonds, B. (2002), "Three Challenges for the Survival of Memetics". Journal of Memetics - Evolutionary Models of Information Transmission", 6 , http://jom- emit.cfpm.org/2002/vol6/edmonds_b_letter.html

Gatherer, D. (2005). "Finding a Niche for Memetics in the 21st Century" Journal of Memetics - Evolutionary Models of Information Transmission, 6. http://jom-emit.cfpm.org/2005/vol9/gatherer_d.html

Miramontes O., Rohani P., (1998) "Intrinsically generated coloured noise in laboratory insect populations", Proc. R. Soc. Lond. B (1998) 265, 785-792

Miramontes O. Rohani P., (2002) "Estimating 1/f scaling exponents from short time-series" Physica D 166, 147–154

Peters E.R, (1991), "Chaos and Order in the Capital Markets", p62 et seq

Puu T., (2003) "Attractors, Bifurcations and Chaos - Nonlinear Phenomena in Economics" , Springer ISBN 3540-402268

Schlather M., Gneiting T., (2001) arXiv: physics/ 0109031 v1 13 "Stochastic models which separate fractal dimension and Hurst effect"

Sprott J.C. (2003), "Chaos and Time-Series Analysis", Oxford ISBN 019 8508 409

Stewart I, (1989) "Does God Play Dice ?", Penguin edn 1990.

US Bureau of Justice Statistics (2007), "Homicide rates from the Vital Statistics", 1900-2002, http://www.ojp.usdoj.gov/bjs/glance/tables/hmrttab.htm

Notes

1. I favour particularly Gretl and use vers 1.60 (2006). [gretl.sourceforge.net/win32/]. Tisean is also pretty good and it could help to run R. In Gretl it is just necessary to type 'hurst' (without quotes) at the command line, having entered a suitable data set which can be done in many ways. But a lot of data is preferable to obtain results and Chebychev

interpolation or alternatively imterpolation by inspection may not be suitable.

2. Ask any keen and dedicated but perhaps nonexpert gardener whether it is easy to outsmart a slug. I think it is not. As for me, slugs are my friends and I get a great deal of pleasure from simply observing them.

3. This does not occur in many standard works, e.g. Sprott (2003) section 9.4.6 p226 which only mention a direct correlation between fractal dimension and Hurst exponent. Given Schlather (2001), for philosophical purposes, which can be relevant here, we may have to probe deeper.

4. Puu (2003) gives a more up to date account than Peters (1991) of economic chaos theory with some cogent observations on economic theory, relating clearly directly to a study of memetics. Basically Puu's book is a book dealing with chaos theory as it can be applied to economics as distinct from Sprott's book which is basically an advanced introductory book on mathematical chaos theory. But unfortunately Puu does not deal with Hurst exponents. Peters (1991) does. Peters (1991) is the classic book in the field, and deals with Hurst exponents in great detail but not as much detail as some would have liked.

Ian Kaplan says in a long useful URL "I thought that the Hurst exponent calculation would be easy .... Sadly things frequently are not as simple as they seem"
http://www.bearcave.com/misl/misl_tech/wavelets/hurst/index.html
I have to say I concur with this as so much of the data seems almost intractable from a Hurst function viewpoint. This is not just equities, which I have also looked at (mainly S. & P. 500 1946 - 2007 over various ranges) but blowflies as well, references as above. (Rohani etc.)

6. Using Gretl (Note 1) there are definite, but in the cases which I have dealt with quite small. differences in starting at a late time for (say a rainfall series - I used recent daily figures for Norman, Oklahoma as these are readily available on the internet - or a S. & P. 500 range) and working backwards or starting from an early time and working forwards, both forward Hurst and reverse Hurst tending to a similar result over a long period of course. But as yet there are no striking results AFAIK and possibly no reason to expect them at this early stage. There is a long way to go yet.

7. Rohani (1998) comments :" fall in the range 0-1 and have intuitive interpretations. A value $0.5 < H =< 1$ indicates what is commonly termed `statistically persistent behaviour'; that is, whatever the past trend in the series,

it is likely to continue in the future, implying a strong degree of predictability. The most extreme case is H = 1 which represents a straight line with a non-zero slope. Here there are no changes along the line when passing from the past to the future; there is absolute predictability in the process. A value $0 <=$ H < 0:5 represents `anti-persistent behaviour': it is expected that whatever the current direction of change, it is unlikely to continue in the future and so predictability decreases. In the limit of H = 0, successive changes in the time-series are totally uncorrelated and prediction is not possible. In summary, white noise is characterized by H = 0, a value of H = 0:5 indicates Brownian motion, and 1/f noise is located in the range

0 < H < 0:5:

The Hurst exponents calculated for Nicholson and Utida's laboratory populations all lie in the range

0 =< H < 0:5.

For Nicholson's blowflies, H is about 0:22 (the first and second halves of the data showed H about 0:46 and H about 0:37, respectively). The bean weevil had H about 0:15, while its parasitoid had H about 0:14: These values lie well within the range expected for a 1/f process. The estimates of H obtained for Utida's populations are, however, quite low. Could these time-series be governed by a white noise process ? To address this question, we used a property of the so-called fractional Brownian motion, generated after performing integration (successive addition) on the time-series generated by uncorrelated Gaussian processes. The integration of a Gaussian uncorrelated signal (with H about 0) produces a random fractal with H about 0:5 (Sprott & Rowlands 1995): We have integrated Utida's data and have found that the of the two integrated signals are H about 1, leading us to conclude that Utida's time-series are extremely unlikely to be Gaussian uncorrelated processes.

There is an important relationship between the value of the Hurst exponent of a time-series and its fractal dimension, D (Feder 1988; Peitgen et al. 1993): D = 2 - H. A straight line with H = 1 has a fractal dimension equal to 1. White noise with H = 0 has a fractal dimension of 2, as expected for a process that is space filling. On the other hand, all the found for the populations above signal that their fractal dimensions are non-integers, as is expected for a dynamical behaviour that has properties of self-similarity.

The foregoing results provide strong evidence that populations free from the influences of environmental forcing can produce fluctuations characterized by well-defined scaling laws. These findings relate to a recent debate regarding the importance of the dominance of high and low frequencies in the power spectra of ecological time-series (Steele 1985; Pimm & Redfearn 1988; Halley

1996; Caswell & Cohen 1995; Cohen 1995; Sugihara 1995, 1996; White et al. 1996a,b; Kaitala & Ranta 1996; Ripa & Lundberg 1996; Sumi et al. 1997; Petchey et al. 1997). Clearly, our results support the generally accepted fact that many natural population fluctuations show 'reddened' spectra, with low frequencies dominant. Traditionally, the dominance of low frequencies in ecological systems has been attributed to external environmental forcing (Steele 1985; Pimm & Redfern 1988; Halley 1996; Sugihara 1996). In contrast, however, we have shown that red noise may arise in laboratory systems, generated by internal population processes in the absence of environmental noise.

What intrinsic ecological mechanisms could be generating these patterns ? It has been shown that simple single-species models exhibiting chaos as a result of strong density dependence may give rise to red noise (Blarer & Doebeli 1996; White et al. 1996b). Whether the demographic parameters required for this are sufficiently realistic as to be expected in nature is a moot point. We propose what we believe to be a more generic mechanism.

We suggest that these dynamics may simply arise naturally from the interaction between (demographic) stochasticity and density dependence. .... The ubiquity of $1/f$ dynamics is one of the major puzzles in contemporary physical science. It is well known that many diverse phenomena generate $1/f$ noise, and many theories been advanced to attempt an identification processes responsible."

7.06 Memetics and the A Series (3)

In "Memetics and the A series (2)" we looked at the Hurst exponent, in an effort to use it in an A series context. Aside from the fact that the use of such concatenations of scalars was simply a tentative start - which could yet bear fruit - there turned out to be a lot of other problems, which were discussed. Rohani's efforts to use the Hurst coefficient for research purposes of not perhaps a pedestrian nature, but at least a conventional one, and the cries of woe of others (not the least Kaplan (Note 4) for example) has made it seem expedient for the moment to consider other avenues as well as the Hurst exponent.

Overall, then, and bearing in mind the difficulties, it was decided to look further, and also to look further at other people's ideas and conjectures. The most obvious thing to consider was the idea that throughout the literature was the 'strange attractor' idea, often applied to memetics. We mention for example R. D. Smith, Gabora, Combs and Rinaldi in this connection.

But first we note that Saperstein (1997) says - and his view merely echoes that of many others -"the paradigm of chaos was intimately associated with battle was certainly well known to von Clausewitz and the earlier Greek military historians. ..... do we gain anything from the visits of the soldier and statesman to the academy of the mathematician and physicist, besides some new, exotic descriptive metaphors (e. g., "strange attractor," "self-organizing criticality")?" ..." It is not evident to me that a single metaphor/tool—like chaos—is available or useful to us in dealing with a world system characterized by "complexity." Instead of specific new tools, these metaphors can contribute to the development of the new attitudes required for the more complex modern world. They can help sharpen minds dulled by a Newtonian world view so as to be alert to all new possibilities." Now whilst Saperstein's article is nicely presented, we are not really interested in pursuing "exotic metaphors"in this context.

We know more or less exactly (in mathematical terms) what we are talking about with the phrase "strange attractor" and that point must be strictly held to when we look at comments embedded in the literature.

That being said we summarise comments on Gabora (Note 1), Combs (Note 2), and Rinaldi (Note 3). Rinaldi seems to benefit us with more detail than most of these writers, but my words in the last paragraph apply, that is we are still not sure what he is talking about in precise mathematical terms.

Notes 1, 2, and 3 aside, out best hope seems still to be with Sprott and Puu if we want to use chaos theory. It is probably plain that I consider the other work above as mainly largely too speculative to take extremely seriously. But the papers listed in "Supplementary References" below, and some of the work cited in Part (2) are (or should be) worthy ones. And even if they are sometimes having little obviously memetically applicable content, they help to build up a picture for further work.

Walter J. Freeman (1991) claims to have found strange attractors in EEG results concerned with the olfactory system, again not too surprising as they are nowadays a simple result of basic maths. He says : "The images suggest that an act of perception consists of an explosive leap of the dynamic system from the "basin" of one chaotic attractor to another; the basin of an attractor is the set of initial conditions from which the system goes into a particular behaviour. The bottom of a bowl would be a basin of attraction for a ball placed anywhere along the sides of the bowl. In our experiments, the basin for each attractor would be defined by the receptor neurons that were activated during training to form the nerve cell assembly. We think the olfactory bulb and cortex maintain many chaotic attractors, one for each odourant an animal or human being can discriminate. Whenever an odourant becomes meaningful in some way, another attractor is added, and all the others undergo slight modification."

He goes on to say "One profound advantage chaos may confer on the brain is that chaotic systems continually produce novel activity patterns. We propose that such patterns are crucial to the development of nerve cell assemblies that differ from established assemblies. More generally, the ability to create activity patterns I may underlie the brain's ability to generate insight and the "trials" of trial and-error problem solving." and "We have found widespread, apparently chaotic behaviour in other parts of the brain." Freeman's theories are currently regarded, in part at least, as unproven and controversial (Bear, 2001)

Lucas (2002) goes on to elaborate the idea into a complete mental (but rather short on detailed maths as presented) scheme but at this point I feel it not necessary to follow suit. By now Tsuda (2001) and many others have also amplified the strange attractor idea into an inclusion in fully fledged brain descriptions.

Well there is certainly not a fully accepted model of the brain as is shown by the multifarious other brain theories available, but it does rather leave the way fairly clear to work out some kind of memetics approach within the A series, and realistically using chaos theory and (perhaps) strange attractors, as we are not bound to find an equation for individual neurons but just general effects, importantly within the A series, and in a way which could also encompass Hurst coefficients.

So if we are to produce a result it may prove best to try to neither achieve a major global nor a highly specific result to start with, like the number of terrorist bombings in a certain area. I did do some sums on such matters but so far, like Edmonds, do not see a very clear way ahead there though I appreciate the work of Gatherer and others.

To recapitulate, we did suggest in an earlier exercise (Yates, 2006) that by suitable primes, it was possible to arrange for a subject to have a particular dream, not a ordinary one but one dependent on future events not considered by the subject in advance. This followed on our earlier dissertations on dream psychology where we noticed that Professor Stickgold could produce dreams, almost to order, in at least some subjects. So the reasoning was, if he could produce dreams after stimulation, would we be able to produce some dreams before stimulation. I also pointed out that there were going to be some knotty problems in interpretation and in repetition of the phenomena. At this point the sort of objections which arose could almost be of the order of those which arose with reference to Asimov's early joke paper on the imagined substance "thiotimoline".

Having made that experiment work on dreaming ahead of the stimulation to dream, I then looked (as indeed I had done before) for factors related to such predictions, which I found even in Professor Hobson's work and again we must carefully point out that he certainly is not of the Freudian or Jungian school, and has always taken a very down to earth view on dreams as can be told from his many books and papers, cited elsewhere in these blogs.

In Part 2 of this paper I outlined the difficulties with experiments on objects and animals. Then the next stage may be to look at memes of a nature which may relate to human beings. There is a large amount of scientific, quasi-scientific, sociological material and so on. For the moment my approach follows that of Sprott to a greater extent than the not dissimilar view of Guastello (Note 5) for various reasons but a principal one seemed to be that it looked relatively easy to me. We leave Freeman's ideas in our pocket for the moment but I will again stress the idea that "the map is not the territory".

From our standpoint, we can refer back to Gale's (1967) excellent anthology of some of the work up to that time, and notice that on p69 he gives three differing ways to deal with McTaggart's paradox these being

(1) The B series alone is enough
(2) The A series alone is enough
(3) either the A series is enough or the B series is enough but they must not be confused with one another.

Now I take a fourth view and that is that we need a B series (which roughly speaking, and exceptions aside) does for physics and an A series which on the whole we use for human perception and other matters, like the possibility of an immortal soul. But both A and B series are required, and often enough they use similar mathematical techniques. If we are not diligent we can certainly fall into philosophical and, depending on the nature of A and B series, mathematical and logical problems. This is a newish view not considered by Gale and other workers. Certainly it would be unwarranted to assume a simple mapping between A and B series as I have indicated in detail in earlier blogs, hence avoiding McTaggart's paradox.

But even within the A series, we are best not generally assuming that "the map is the territory" and if we are careful there is no need for an infinite regress. To put it very simply, we see things we think we know should be found in an A series,we try to identify them (as we did with the Hurst exponent in "Memetics and the A Series (2)" for example) and then we describe them in terms of the A series. This of course normally may require mathematics but we have to be careful that a past, present and future are involved. This will give us an A series or a pseudo A series written after the form of a B series at worst.

It's very hard to know how far we can go with these pseudo A series in my opinion.

One thing that can be done is to try to extend non-linear dynamical psychology. First we note two very interesting recent cases of that by Sprott (2004, 2005). These involves such things as a mathematical description of Love and Happiness and this is really quite a clear mathematical example to the point where it is used by Strogatz (1994), Speigelman (1997), Parwani (2001) and others to provide elementary/advanced training courses in chaos theory, catastrophe theory and the like. Somewhat similar work has even been done on the Mexican wave (Farkas, 2006).

Most of this is welcome enough, it seems to me. But we need more useful attempts to universalise the phenomena and specific figures for actual events are often lacking, although some figures appear in the work of, for example, Jones (1995) , Helbing (2000) and Aks & Sprott (2003). The latter paper is interesting for two reasons : Firstly it goes into much detail on the Necker cube phenomenon (which of course is like the duckrabbit discussed in the earlier blogs) and secondly insofar as it applies directly to neuroscience and we are left musing the comment "During the last decade, it has been established that a large number of natural systems containing several interacting individual components have statistically similar dynamical properties, independent of the particular details of the system (italics mine). Examples include earthquakes, population dynamics, DNA base sequence structure, epidemic outbreaks, and various cognitive and reaction time behaviours. Examination of the statistical properties of these system fluctuations has revealed dynamics with well-defined generic scaling properties in the form of power laws .... etc". We bear in mind the Packard-Ruelle-Takens method and also Aaronson's (2006) recent comments about the universality of NP-complete results, considered since the early 1970s.

But the work of Aaronson (2006) makes it clear where conjecture can lead simply to disaster and at least in his flamboyant exercise in that paper, he clearly seems right into the B series when he speaks of quantum computing and there are problems there, when serious attempts are made to express directly in terms of the B series, ideas that in fact seem to come from the A series.

Strogatz (1994), Sprott (2004, 2005), Jones (1995) Spiegelman (1997) and Parwani (2001) seem to be writing largely in A series terms but flowing over to a limited extent to the B series, whereas the problems in this sort of position come out clearer in Aaronson's (2006) work I think. Aaronson takes a great interest in quantum computation so this effect seems normal, natural and almost expected. In fact Aaronson's (2006) exercise in places looks a bit like one of Sprott's (2004, 2005) exercises on love and happiness gone wrong, perhaps with some humorous intent by Aaronson but it seems to me close to the amusement often allied with bizarre and important curiosity. One of my own pet slugs seemed to show such amazed curiosity when I put down extra food for it last night, but perhaps I anthropomorphise there a bit too much.

We need to bear in mind that some, if not all, of these A series uses can be construed as faux-A series. That is in the sense that some people would claim to find similar B series results in one way or another. But we must bear in mind that we have tried to find things that we know we would expect to find in an A series, identify them, describe them and try to use them. And we really need to distinguish two things from one another. Firstly, is there a difference between the A series and the B series ? Secondly, are certain items, for sure, A series or B series? Now we can be sure on the first query. The A series gives us a past, a present and a future and more, while the B series perhaps gives easier mathematical conjectures. Roughly like the difference between tensed time and tenseless time. There is a difference between A series and B series. On the second query, usual physical 'rough assumptions' (no worse than say renormalisation problems) can always appear at the present state of the art. Certainly some items or manipulations clearly seem one or the other but as elsewhere in life there are margins of doubt. At the moment we are probably at least no better off than physics was on mass renormalization without the RG groups. We do know, though, that an inappropriate mix of A and B series will lead to paradoxes according to McTaggart.

Well now we have indicated at least some of the ways the A series may be used. There definitely seems the groundwork for further progress now.

For example there is beginning to appear ways, at least in principle, how an element of precognition could be detected. Typically dream precognition has been said to be related to intense emotional experience. It has to be stressed that at the current state of the art, no seriously satisfactory evidence other than anecdotal is available and an open and rather sceptical viewpoint has to be considered for that, but of course there are countless anecdotal cases which claim precognitive results in cases like 9/11 .

In the case of the experiments that we have done, it is true that we obtained positive results. But we would certainly like to obtain a series of independently controlled trials to the same effect. With direct mental involvement particularly, this may not be easy. The feeling is that usually results evaporate when supervision and scaled up psychology experiments are tried for anything so ambitious. The earlier work in this blog has tried to find out why positive results even for very large and clear systems take so long to emerge. And more often than that, these involved simple models like the solar system, physical rather than obviously psychological. Stanley Milgram, whose work I greatly admire, was one of those who was able to stand by his results in experimental psychology, to world acclaim. On the other hand some effects, even when they are physically obvious on examination, may simply remain as 'anecdotal' and be suppressed in the way results regarded as 'currently unfavorable' so often are. For very straightforward experiments sometimes this is clearly this suppression is often dispensed with quickly. We look for example at early repetitions of the Millikan oil drop experiment where some results were clearly suppressed. And we earlier in this blog mentioned chemical reactions, where results dependent on chaos were simply apparently omitted as not falling within the limits which experimenters would have expected. And then of course we have the almost-parable of Semmelweiss.

Here we are not witch-hunting in advance - but we bear in mind that many of the early Rhine experiments are now regarded as fraudulent after clear admissions by their practitioners and other experiments on children as apparently involving pedophilia also have perhaps to be excluded, though that it seems that 'pedophile' is now a current cry by small children resisting arrest by the police,whether their complaints are true or not. I'm all in favour of doubt to the level of people like Randi but we do need to be sure if and when it is justified given the circumstances.

We looked at somewhat dubious and borderline work such as some experiments on 'synaesthesia' and earlier in this blog I have mildly queried some examples, done interviews of synaesthetes and experiments on them and tried to estimate roughly the apparent present state of the art in that field. The Perky effect appears also to be borderline but the evidence is still not all in, though some hazard a guess that most of it is. And now we are left with the matter of near death experience. I have spoken to Dr. Peter Fenwick about this and he seems to hold a very positive point of view, and Dr. Bruce Greyson also suggests that there may be much evidence available which has not yet come in but allows that consideration of NDEs may also have a psychotherapuetic value. Now I have unfortunately in my blogs made things a little more difficult as I have pointed out that hitherto considered but then omitted cases of NDE may actually be part of evidence of a striking success in discovery of NDEs !

I do not ignore the many apparent continuing successes in ESP and such like and certainly Dr. Fenwick mentioned some of those to me. However other people, often very eminent ones have very negative opinions on ESP and such like. At the moment therefore, I feel that it may be contentious to include these issues in a program. Dean Radin's (2007) work, for example, is well known but I enclose a current Wiki URL on it which is very negative at time of writing, and illustrates the problems which anything unusual tends to have, often for good reason.

Experience here suggests certainly with dream precognition that some direct involvement with brain action, after the precognition, possibly occurs, if there is any precognition which I am not prepared to gamble on. The cases described here of course simply used something like the computer game technique pioneered by Stickgold and mentioned above, but that clearly involves a lot of brain action. In many of these cases it is so easy to set up a large series of controlled experiments and I am not yet convinced as to how they would be best best operated. It would not have been good science, for example to try to claim that nobody can carry out extremely rapid mental calculations and in doing so ignore Ramunjan, Erdos or others in the small class of savants. So using the mathematical techniques used in this article as above, exemplified typically by Sprott (2004, 2005) a more detailed presentation may develop, perhaps in later blogs of this series.

I would suggest that there is still a good deal of work to be done even on very simple cases like dream precognition which seems to do no more than show some A series influence, and near death experience which may do the same and for the moment these seem like the most likely success candidates.

The "rotters and scoundrels" in the case of the existence and relevance of the A series, happily seem to be conventional physicists as everyone else would seem to see that we do each have a past, a present and a future, and that some people do believe in free will and God. To not admit that can lie on the borders of solipsism, and even madness, and if physicists persist in their views we can draw our own conclusions. But it could be argued that the jury is still out on contrived dream precognition.

References

Combs A., (1995), "Chaotic and Strangely Attractive", Dynamical Psychology. (journal) also in "Mind in Time: the Dynamics of Thought, Reality and Consciousness", co-edited by Allan Combs, Ben Goertzel and Mark Germine ; and at www.goertzel.org/dynapsyc/1995/combs.html

Gabora L., (2001), "Cognitive Mechanisms Underlying the Origin and Evolution of Culture", Doctoral Thesis, Free University of Brussels.; and many other publications with Aerts D., etc quoted at http://www.vub.ac.be/CLEA/liane/Publications.htm

Gabora L., (2007) A case for applying an abstracted quantum formalism to cognition. Invited paper for New Ideas in Psychology quant-ph/0404068

Lucas C., "Evolving an Integral Ecology of Mind", http://www.calresco.org/lucas/eiem.htm#refs

Radin D. (2007) http://en.wikipedia.org/wiki/Dean_Radin ; these wikis alter and accesses after Jan 2007 may differ from the case chosen for illustration.

Rinaldi, S., G. Feichtinger and S. Wirl (1994) 'Corruption Dynamics in Democratic Systems', Research Report #168, Institute for Econometrics, Operations Research and Systems Theory. Technical University of Vienna. Vienna, Austria.

Rosicky A., Pavlícek A., (2006), "Knowledge and Conceptual Information", Proceedings of the 50th Annual Meeting of the ISSS, ISSS 2006 Papers

Saperstein A., (1007), "Complexity, Chaos, and National Security Policy: Metaphors or Tools?" in "Complexity, Global Politics, and National Security" edited by David S. Alberts D.S. , Czerwinski T.J. , National Defense University Washington, D.C.

Smith, R. D. (1998) 'Social Structures and Chaos Theory' Sociological Research Online, vol. 3, no. 1, http://www.socresonline.org.uk/socresonline/3/1/11.html

Supplementary References

Aaronson S., (2006), "Notes for a talk given at the Stanford Institute for Theoretical Physics, December 15, 2006 ", http://www.scottaaronson.com/talks/anthropic.html

Aks, D.J. & Sprott, J. C. (2003) Resolving perceptual ambiguity in the Necker Cube: A dynamical systems approach. Journal of Non-linear Dynamics in Psychology & the Life Sciences, 7(2) 159-178. Helbing D. et al , (2000), "Nature", (407), 487

Bear M.E., Connors B.W, Paradiso M.A., (2001), "Neuroscience", p612.

Earn D.J.D., Levin S.A., Rohani P., (2000), Science, "Coherence and Conservation" November, Vol 290, 1360

Farcas I.J., Vicsek T., (2006), arXiv: physics/ 0601181 v1 23 Jan 2006

Freeman W. J., (1991), "The Physiology of Perception", Scientific American, Vol 264, (2) Pgs. 78-85.

Freeman W. J., (1999), "How Brains make up their Minds", p104 and elsewhere, Phoenix, ISBN 075381 0689

Gale R.M., (1967), "A Philosophy of Time", MacMillan, London WC2.

Guastello S.J., (2006), Nonlinear Dynamics, Psychology, and Life Sciences, Vol. 11, No. 1, pp 167-182.

Guastello S.J., Pincus D, Gunderson PR (2006), Nonlinear Dynamics Psychol Life Sci 2006 Jul; 10(3) :365-99.

Jones, F.J. (1995). The Structure of Petrarch's Canzoniere: A Chronological, Psychological and Stylistic Analysis. Cambridge, England: Boydell & Brewer, www.boydell.co.uk/1764.HTM

Keeling M.J., Rohani P., (2002), "Estimating spatial coupling in epidemiological systems: a mechanistic approach", Ecology Letters, 5: 20-29

Keeling M.J., Rohani P., Grenfell B.T., (2001) "Seasonally forced disease dynamics explored as switching between attractors" Physica D, 148, 317–335

McSharry P.E., (2005) "The Danger of Wishing for Chaos", Nonlinear dynamics, psychology, and life sciences,9(4):375-397

Miramontes O., Rohani P., (2002) "Estimating 1/f ^alpha scaling exponents from short time-series", Physica D 166 147–154

Parwani R., (2001), "Complexity", http://staff.science.nus.edu.sg/~parwani/c1/book.html

Pelletier J.D., (1995) "A Stochastic Diffusion Model of Climate Change" arXiv: ao- sci/ 9510001 v15

Rinaldi S., (1998), "Laura and Petrarch: An intriguing case of cyclical love dynamics", Siam J. Appl. Math., Vol. 58, No. 4, pp. 1205 - 1221

Rohani P., Green C. J., Mantilla-Beniers N. B,. Grenfell B. T., (2003), "Ecological interference between fatal diseases" Nature, Vol 422, 885

Rohani P., Wearing H. J., Cameron T., Sait S. M., (2003) "Ideas and Perspectives: Natural enemy specialization and the period of population cycles" Ecology Letters, 6: 381–384

Rohani P., Keeling M.J., Grenfell B. T., (2002) "The Interplay between Determinism and Stochasticity in Childhood Diseases" vol. 159, no. 5 The American Naturalist (May)

Spiegelman M., (1997), "An Introduction to Dynamical Systems and Chaos", http://www.ldeo.columbia.edu/~mspieg/Complexity/Problems/

Sprott J.C., (2004) "Dynamical Models of Love", Nonlinear Dynamics, Psychology, and Life Sciences, Vol. 8, No. 3, July.

Sprott J.C., (2005), "Dynamical Models of Happiness", Nonlinear Dynamics, Psychology, and Life Sciences, Vol. 9, No. 1, January.

Strogatz, S.H. (1994). Nonlinear Dynamics and Chaos: With Applications to Physics, Biology, Chemistry, and Engineering. Reading, Mass.: Addison-Wesley.

Tsuda I, (2001), "Toward an interpretation of dynamic neural activity in terms of chaotic dynamical systems", Behavioral and Brain Sciences , 24, 793–847

Wearing H.J., Sherratt J.A., (2001) "Nonlinear analysis of juxtacrine patterns", SIAM J. Appl. Math. 62, 283-309. ; others on http://www.ma.hw.ac.uk/~jas/researchinterests/juxtacrine.html

Yates J., (2006), "Can dreams predict the future ?". http://ttjohn.blogspot.com/2006/04/do-we-dream-of-future.html

Notes

1. Gabora (2001) "Abstractions are not only driven by the memetic fitness landscape, they feed back on and actually alter its topology. Much as the evolution of rabbits created ecological niches for species that eat them and parasitize them, the invention of cars created cultural niches for gas stations, seat belts, and garage door openers. As one progresses from infant-hood to maturity, and simple needs give way to increasingly complex needs, the trajectory of a stream of thought acquires the properties of a chaotic or strange attractor. The landscape is fractal (i.e., there is statistical similarity under change of scale) in that the satisfaction of one need creates other needs. This is analogous to the fractal distributions of species and vegetation patterns described by ecologists (Mandelbrot[60], Palmer[69], Scheuring & Riedi[83])."

Now this sort of thing is all very well but there seems to be no clear reason why such an analogy may have direct mathematical meaning. And I further point out that attempts are made to drag quantum mechanics, still a useful buzz-word, into all this in Gabora (2007). I would say that whilst this may be a brave attempt at interdisciplinary study, I have expunged similar work from my files since the days of G.D. Wasserman and his field theory about ESP.

238

Call me a sceptic if you like, but sometimes I can understand why David Deutsch seems to get upset by people like Brian Josephson, who really should know better, possibly unlike Gabora who is really just having a rather flashy try. And hard scientists can ponder (a little bit) on her stuff - should they choose.

2. Rosicky (2006) says: "Combs [2002] explains cognitive/mental processes by term (chaotic) attractor – processes on the edge of chaos reach over two dynamic stable states - Lorenz's strange attractor." The rest of the paper is not completely clear. So let's try Combs. Combs (1995) says "To my mind Tart's ideas are of the first order, but can benefit from more recent advances in the sciences of complexity, which yield more dynamic and fluid conceptions of the nature of systems. For example, a state of consciousness can be reconceptualized as an attractor. Speaking informally, an attractor is a condition to which a system is drawn by its own nature. If a cup is placed slightly tilted on a table, it will roll about in a spiral till it comes to rest standing up. This latter condition is termed a static attractor, because it represents the static position to which the cup is disposed. More interesting are cyclic or fixed cycle attractors. The human heart, for instance, runs through its cycle many times each minute. The moon passes through its various phases each month. These, and many others, are instances of systems that naturally settle into predictable cyclic routines. Most interesting, however, are the class of attractors that are neither fixed nor precisely predictable. These are termed strange or chaotic attractors." As for Charles Tart, you can get opinions on his work from http://skepdic.com/tart.html . So Combs uses the ideas of Tart - not necessarily all incorrect - to suggest that a state of consciousness is a chaotic attractor, bringing in many elements to coalesce to a unified pattern.

Combs may be on better ground when he brings William James's ideas into it and suggests consciousness is "a constantly changing process, clearly not static or even following a fixed cycle, but nevertheless one that has an identifiable global character, at least for each individual. Memories come and go, thoughts pass through the mind only to disappear and return again later, moods are continually changing, and alertness and energy levels vary from hour to hour. These are the elements of a kind of mental soup, or more accurately a kind of mental weather, with the equivalent to the latter's constantly fluctuating temperature, humidity, wind, barometric pressure, and so on".

So in essence Combs is saying that, like the weather, consciousness may be represented by mathematical equations and some of these may arise from chaos theory. Now many simple equations need chaos theory to represent them, as Sprott's book (ref in Part 2) makes clear. And it is clearly fair to say that a chaotic attractor may thus occur in some of them, equations being what they are.

Combs then says:"It is not surprising that weather is chaotic. Indeed, the elements that comprise it, such as temperature, oscillate in an identifiable cycle from day to day, but cannot be predicted with precision. What is more, it is unlikely that temperature fluctuations ever follow exactly the same course on any two days. Much the same can be said about mental weather. It is formed of the interaction of elements such as moods, thoughts, memories, and so on. These are Tart's original psychological functions. For some, such as moods, there is already empirical evidence that they are chaotic (e.g., Combs, 1995; Daley, 1994; Sacks, 1973/1990; Winkler & Combs, 1993), while virtually all are consistent with the general description, above, of chaotic processes. As a group, their interaction, like the interaction of the elements of the weather, yield an exquisitely complex process fabric that we know as consciousness. This fabric is far too complicated to describe in detail, but efforts have been made to mathematically conceptualize it as a grand chaotic attractor."

Apparently consciousness, on such models, is to hard to conceptualise in detail. But Combs perseveres "If the system gets stuck in the attractor of a wrong solution, subsequent recall will be incorrect. What is needed is a process that keeps it from settling down too quickly in the first attractor basin that comes along. This process is chaos. One can easily think of it as operating in a similar fashion during the search for a solution to a mathematical or linguistic problem, or a quest for the right artistic expression. Chaos is the antidote to stasis and stagnation."

But we still need a detailed mathematical description of what is going on. As such writers do, Combs says much more, in various places, but that is a brief gist of it. Rinaldi perhaps enlightens us with more detail which we describe briefly in Note 3.

3. R.D. Smith (1998) is good enough to enlighten us as to Rinaldi's (1994) thoughts: "In order to have a formal notion of structure which is consistent with the chaos theory paradigm we now need to return to a requirement so far unelaborated. In its most precise rendering chaos can only arise when the possibility of any given state repeating itself is potentially zero. To take the illustration of a strange attractor such as the Lorenz attractor (as in Figure 4) what is needed is a situation in which the orbital pathway of a flow or flux can

continue for an indefinitely long period of time (for eternity) without ever passing through the same point twice. If this condition is not met then the orbit is not in fact chaotic but periodic even though highly convoluted. What this in turn means is that the phase-space in which the flux is propagated should be continuous and not quantised. A quantised space (says Smith), however large, is effectively finite and thus cannot provide for truly chaotic behaviour. Other more deductive approaches to large-scale pattern identification can also be used. Recent hypothetical studies by have produced the following application of 'strange attractor' concepts to the flux of political popularity...

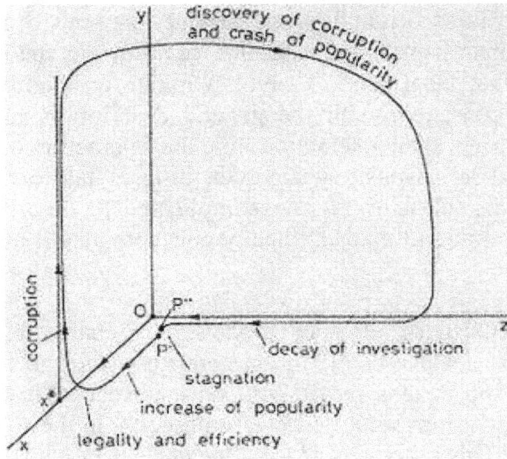

and follow this with a brief mathematical description also due to Rinaldi. Smith also mentions the duck/rabbit example of Wittgenstein, somewhat similar to the spotted dog against a spotted background, which is neverthless easily distinguished by the human eye. To his great credit Smith also says:" Chaos, like Relativity, Darwinism and Mechanism before it, has the potential to be transformed into a metaphor and to have its terminology misunderstood and misapplied". And that is what we must be sure not to do here.

4. Ian Kaplan says in a long useful URL "I thought that the Hurst exponent calculation would be easy....Sadly things frequently are not as simple as they seem" http://www.bearcave.com/misl/misl_tech/wavelets/hurst/index.html

5. Guastello says "First, we should not become overly preoccupied with its (NDSs) [nonlinear dynamical systems] principles of system connectedness at

the expense of forgetting the basics – attractors, bifurcations, chaos, fractals, self-organization, catastrophes, and so on"

"Before basic NDS, change was understood in the social sciences as only one amorphous entity –change."

"It is also historically important, in my opinion, that two of our most central concepts, chaos and fractals, originated about a century ago when aviation was in its infancy and the very first papers on rocket science, not to mention the Theory of Relativity, were being published and discussed. In our travels we picked up nonlinear topology, information and entropy concepts, catastrophes, and self- organization; by the mid-1980s had begun to see formal connections among all these systems phenomena. Given the time horizon involved, it is doubtful that post-modern philosophy can claim with a straight face any more credit for the scientific developments than perhaps generating a little more interest than what would have been the case otherwise. It is also doubtful that any philosophical genre could claim more credit than any other genre for having discovered creativity itself."

In my opinion this is very true and by using McTaggart's paradox I am trying now to rid us of the Tychonic illusion and present us with time travel and immortality.

Guastello then goes on to say "where it has been possible to compare the accuracy of nonlinear and linear models, and the nonlinear model was adopted as the conclusion, the average ratio of variance accounted for was 2:1 in favor of the nonlinear model (Guastello, 1995, 2002). This is obviously a utilitarian criterion of success."

Fair enough. and here I must quote one of the many papers in the field by McSharry (2005) as to likely problems (but which also offers some methods) which have occurred in the past to these methods, and may occur when applied here.

" With the discovery of chaos came the hope of finding simple models that would be capable of explaining complex phenomena. Numerous papers claimed to find low-dimensional chaos in a number of areas ranging from the brain to the stockmarket. Years later, many of these claims have been disproved and the fantastic hopes pinned on chaos have been toned down as research with more realistic objectives follows." etc etc

242

Guastello's approach is outlined in Guastello (2006) and is well worth considering but there are difficulties in its complexity of arrangements

7.07 A simple new approach involving application of category theory to a number of obscure results and hard to fathom facts.

1. There may not be enough computational ability in the brain to explain some phenomena, including even widely accepted ones like intelligence and insight. (note 3)

2. A quantum computer effect in the brain is frequently postulated nowadays and that would certainly allow extra computing ability. There is no commonly acceptable way this can happen yet except for MWI (Many worlds interpretation). So we take that to be a semiplausible working hypothesis.

3. The work of Hosten and of Vaidman suggests these worlds may in some sense be 'real'. (Dowling, 2006)

4. We also have McTaggart's paradox which implies that time requires an A series and a B series. This does not infer that obviously quantum effects, in the sense of Penrose's microtubules for example, help to describe normal brain activity. This is all to the good from the Occam's razor (parsimony principle) standpoint. We are just describing things as we believe them to be.

5. Now we find that dreams can be an indicator of the future, as in one of my preprints under consideration for publication (Yates 2006). This does not imply that we can get racing results etc. from dreams in some way, but simply is a reasonable extension of current mental processes in which we can imagine both past and future, but feel located in neither, though in dreams we seem to be floating a little freer for whatever reason, and my view is closer to that of Hobson than to many others. I have nonetheless extended Stickgold's work a little, perhaps allowing us to dabble in the A series.

6. Because of the obvious relevance of synaesthesia to functionalism as pointed out by Jeffrey Gray and obviously even more relevance to computationalism, as referred to in Yates, (2006b) I tried to get some more clear cut synaesthesia results but these are still pretty elusive or even a sort of

243

Perky effect style failure as Yates (2005) has pointed out for some time. I figured that if available, this could give a better idea of the exact nature of the A series. Be that as it may, there seem to be quite a number of other reasons why we cannot simply assume computationalism (Notes 1, 4) and this can make it far more difficult to set up an adequate A series.

7. So what is really wanted now is something which gives solid physical prospects, such as more detailed dream experiments (as I tried to point out in Yates (2006)) or OBEs (out of body experiences) or NDEs (near death experiences). The mathematical and physical prospect of many worlds is better than those of much of today's physics, in fact a useful quantum computer is likely to be built by 2020 (Dowling, 2006). As these things go, the present supposed restrictions of any MWI is certainly likely to evolve in that time, and of course we have as yet no details as to how. The A series for instance could be a proper class (note 2) and to begin with we may have to map a pseudo A series onto a mock B series to get results, and in effect I recently suggested something like that in my blog, Yates (2006c). The 'block universes' of the B series type have been relatively easy to handle so far, though philosophically and to the intellect not altogether satisfactory, without an A series.

8. I have looked briefly at the recent NDE work of Dr Peter Fenwick. For example in Fenwick (2004) "The flat electroencephalogram (EEG), indicating no brain activity during cardiac arrest, and the high incidence of brain damage afterwards both point to the conclusion that the unconsciousness in cardiac arrest is total. You cannot argue that there are "bits" of the brain that are functioning; there are not." It is about that time that a NDE or OBE sometimes occurs. Now the glib idea would be that the reason is that the person is somewhere fully alive in one of the other worlds in the MWI. Why? Perhaps because, as we have had to assume already, the computing power of the brain must be spread over the many worlds and 'our' world has temporarily dropped its bundle so some of the worlds are keeping things together. This may arise from a revised interpretation of Vaidman and Hosten but I certainly do not think things are quite that simple and I can visualise some objections of MW enthusiasts. Still, there is added ground to consider here, given MWI and it seems to me that it may be possible to build on the A series given more data, and most importantly to exclude irrelevancies and reasons for the observed OBE and NDE whilst doing so. But we would be better with a lot more OBE and NDE results if we can get them. And we are certainly getting some dream results with the A series approach but further progress is difficult. These matters are quite important as so far a general assumption is that death is a very simple end to life, like the tail of a snakelike object stretched out along

the t axis means the final end of the snake in the block universe. A simple B series interpretation of MWI could do just that as the snake could then simply continue (or not) in other distinctly different of the MW which are diverse from 'our' world. McTaggart's paradox and tensed/tenseless time place the matter in a different light. No-one wants to get the truth about death wrong, whatever the facts are.

References

P., (2006), 'Oxford University reckon that "A useful (quantum) computer by 2020 is realistic,"' Nature 440, 398-401

Dowling J.P, (2006), Nature 439 (Feb 23):919-920.

Fenwick P., (2004), "Science and Spirituality: A Challenge for the 21st Century"

Gray, J (2004) "Creeping up on the Hard Problem" , Oxford University Press.

Yates J., (2006), "Can dreams predict the future ?"

Yates J., (2006b), ""The Application of McTaggart's results to Consciousness Studies and Category Theory." and Appendix A below.

Yates (2006c). "Preliminary plans for a detailed MacA" Yates J. (2005) unpublished notes, available.

Notes

1. Other reasons can even be simply practical, as for example as illustrated in the difficulties which seem to be encountered with purely computerised CBT (cognitive behavioural therapy) without a trainer/instructor and I also refer later to, for example, the everybody's unsatisfactory Loebner prize results.

2. Goldblatt's 'Topoi' refers to a 'proper class' as 'a class which is not a set'.

3. I am interested in pursuing and possibly considering modifying or improving on any NDE and OOB results believed new or important. The reasons are very briefly as follows.

Time Travel in Theory and Practice

There may not be enough computational ability within the brain to achieve the results which it produces - opinions differ and some feel some form of mental 'compaction', as yet not understood, is how it works or a way like Penrose's microtubules idea. (The latter idea by the way I consider totally wrong.) The recent work of Hosten suggests the real possibility of many alternative actual worlds, some very similar to this one, in which computations are being performed for us. It seems that something very strange is going on, is that it is even claimed that a computer has obtained results without actually being run in this universe, and that could imply that it is being run in other universes. I accept that other interpretation's of Hosten's results are possible but it is not even clear that a modified version of Bohm's interpretation may effectively get rid of these worlds. But making the obvious assumption- these other worlds pro tem can be assumed to exist - (and of course a varied but not incongruous result may occur even if they do not, perhaps along Bohmian lines for example) means that these worlds interact with the present world, in some fashion. How? Well some may say in a purely abstract fashion, as if they simply involved imaginary vectors. But our A and B series approach looks as if it may allow something which seems almost like a real (rather than in some mathematical sense virtual) interaction, and in fact already psychological experiments have been done and sent for publication along lines which look a little like precognition, though not in quite the way most people would conceive it. In fact we are almost as iconoclastic as Susan Blackmore and accept that many of her studies may echo the norm. But we need more experimental results, hopefully, to see how much better a picture we can obtain and any new results which throw light on out-of-body experience or near death experiences could help with this. The present work does not depend on any changes in quantum theory or, seemingly, even unconventional use of quantum theory and in fact so far our techniques fit in with classical physics, even non-relativistic physics where that would apply. However modern mathematical category theory does seem to be necessary, but we do have to try to squarely face the problems that immediately seem to make many category theorists descend to one or another variety of computationalism. Specifically in this regard, qualia problems are always uppermost in my mind. As well as this there are many factors still to be borne in mind, e.g. the unlikelihood of the strict Loebner prize test requirements being met in the near future, and that is always the stance.

4. I do not like to refer to science fiction in a scientific work as even where relevant it often grossly oversimplifies and can simply be incorrect, but in places it is by now coming so close to perilous likelihood that it ill behooves

anyone to ignore the dangers of computationalism. Global warming is so high that large commercial interests and governments are likely to consider any alternative to avoid their own corruption being blamed for world resource exhaustion and far worse. So, to illustrate the physical dangers and moral issues, I refer here to a quotation from science fiction, (used in another context about functionalism by Russell Standish; it is even more relevant to computationalism.). As Douglas Adams writes about Arthur Dent's brain: "It could always be replaced," said Benji reasonably, if you think it's important." Yes, an electronic brain," said Frankie, "a simple one would suffice." "A simple one!" wailed Arthur. "Yeah," said Zaphod with a sudden evil grin, "you'd just have to program it to say What? and I don't understand and Where's the tea? - who'd know the difference?" "What?" cried Arthur, backing away still further. "See what I mean?" said Zaphod and howled with pain because of something that Trillian did at that moment. "I'd notice the difference," said Arthur. "No you wouldn't," said Frankie mouse, "you'd be programmed not to."

Appendix A

Addition to "The Application of McTaggart's results to Consciousness Studies and Category Theory."

To save words, I point out the paradox of J.M.E. McTaggart, and the related topics of tensed and tenseless time, are still very current matters. This is not a review so there is no need to refer to the many current papers on the latter.

The semantic arguments embodied in such work as Boroditsky (2000) and which run through much other work, interestingly enough, make clear the importance of psychological experimentation for the achievement of specific results in research on time, and I certainly concur with that. In fact to do so, is part of the present investigation. In that sense, mentioning the semantic aspects of the work is of real value as myself, Boroditsky and others, are already picking up on this point. The idea of McTaggart's paradox being 'out of date', as one reviewer in effect brusquely informed me, is of course another matter.

That would be as naive as to suggest that one popular version of Zeno's paradox is meaningless as a tortoise does not run faster than Achilles. In that case, clearly the author was making a number of mathematical points, and even today, modern results such as those of Kwiat and Hosten can actually imply, in the MWI, that the Zeno paradox could even lead to many real worlds other than the one we know! So it is wrong to write off prematurely any problems in a paradox or pseudo-paradox, simply as semantics, if we feel like it. Perhaps to abuse semantics in this fashion is to create the worst kind of argumentum ad hominem. What we do is almost the opposite to writing off semantics, or saying it is not with us. What we are discussing here is a universe which can perhaps best be described in A series and B series terms, hence some problems. At the present state of play, I actually take the view that McTaggart should perhaps have curbed his enthusiasm somewhat in some parts in his important work, as for example in his way of involvement of 'God', but the enigma certainly stands. Equally, semanticists and computationalists should curb their enthusiasm at the power of their methods. These methods are powerful, I like them too, but they are no 'silver bullet' for as yet unexplored paradoxes and other situations. That way leads to stultification of all philosophy and thought.

References to Appendix A

Boroditsky L., (2000), Cognition 75, 1-28

Dowling J.P, (2006), Nature 439 (Feb 23):919-920.

7.08 Category theory applied to a radically new but logically essential description of time and space

Abstract

McTaggart's ideas on the unreality of time as expressed in "The Nature of Existence" have retained great interest for many years for scholars, academics and other philosophers. In this essay, there is a brief discussion which mentions some of the high points of this philosophical interest, and goes on to apply his ideas to modern physics and neuroscience. It does not discuss McTaggart's C and D series, but does emphasise how the use of derived versions of both his A and B series can be of great virtue in discussing both the abstract physics of time, and the present and future importance of McTaggart's ideas to the subject of time. Indeed an experiment using human volunteers and dynamic systems modelling which was carried out is described, which illustrates this fact. The Many Bubble Interpretation, which also derives from McTaggart's ideas, is discussed and various examples of its use and effectiveness are referred to. The Schrodinger Cat paradox is essentially resolved in principle, the effect interpretable, Kwiat's recent result referred to, and the newly discovered described.

Introduction

I began in the late nineteen sixties , with Professor R.O. Gandy in Manchester, England, by trying to describe and attempting to completely incorporate into a mathematical system, the laws of physics. I used basic methods, such as those of Gentzen, Heyting Brouwer etc., etc. But both I and Professor Gandy found a practical solution, even in the very abstract, to be too difficult at the time. I believe we both thought that we needed some new mathematics, which either did not seem to exist or which we simply had not located !

Now the work of Turing, and later Chaitin and Connes, for example, should have helped but somehow it seemed to me necessary to go even deeper down and more basic. In fact the philosophy of approach with which I began was that of the early formal system theory of Smullyan (1961). Clearly on the face of it, it looked as if strange mathematical constructs like that of Godel universes as well, could be included in such an approach. But at that point, the pieces did not seem to fit. For example, pursuing the Turing path, which has been trod by so many workers by now, like for example Juergen Schmidhuber, was not going to be enough. There was more to it than simple computability problems, we needed to go in a sense to a higher level. Even using the physically peculiar looking results of quantum theory which have by now been incorporated into modern methods of quantum computing, and some of the early results of which, for example, were first published in a journal which I founded and of which I was Editor in Chief for many years (Feynman, 1982) could certainly enlighten us and might well have to be included in some more complete description of the universe which more finally became of use to us, were probably too intellectually ad hoc and thus too flimsy to effectively suppress or even mollify the deep angst of our lack of basic understanding. It was almost like trying to understand modern number theory in a position where transfinite numbers had not been invented. There almost had to be "another dimension or dimensions", or even another "kind of dimension".

Early string theory was around at the time, but at this very basic level, the explanation was unlikely to have that kind of simplicity. It was likely to be much more basic, deep and profound. In sum we were looking for fundamental mathematics, not just the simple technical physics that string theory, even today, would seem to amount to.

I felt in the early nineteen sixties, and still feel now, that the great Emmy Noether, who has since been described flamboyantly but possibly realistically as the greatest mathematician who ever lived, in making a comment on the equality of numbers outlined a more basically sensible approach and that comment should be able to enlighten our understanding. "If one proves the equality of two numbers a and b by showing first that "a is less than or equal to b" and then "a is greater than or equal to b", it is unfair, one should instead show that they are really equal by disclosing the inner ground for their equality". The same idea applies of course if, for example, a does not equal b but the formulation of our problem here is thornier. And although I end up here talking about the A and the B series, it is not with the idea of using a simple logical, physical, or mathematical proof but a striving for something closer to the absolute.

McTaggart and Angst

Referring now back to space and time, Buber (1959) pointed out 'A necessity I could not understand swept over me: I had to try again and again to imagine the edge of space, or its edgelessness, time with a beginning and an end or a time without beginning or end, and both were equally impossible, equally hopeless – yet there seemed to be only the choice between the one or the other absurdity'. The problem here is that when Buber tried to get down to philosophical details he just had not got the right stuff and relativity theory shows us that. There is really no certain reason, using relativity, why time or space would have a beginning or an end - philosophical problem solved.

Now we could say that Buber's confusion was caused by his acceptance of Newton's concept of space rather than Leibniz's. In Newton's world-view physical objects could exist by being in space, but space could exist even if devoid of any physical objects. In Leibniz's view, objects existed anyway and could touch one another, be separated by various distances and so on but space, per se, did not exist. This immediately resolved Buber's problem. One can solve such a problem by showing that it contains an untenable proposition. In this case the problem was not with space itself, but with Newton's conception of space. The answer was to accept Leibniz's more economical view, or simply to look for a consistent definition of space, which without relativity was hard to find.

250

Time Travel in Theory and Practice

McTaggart (1927) reasonably showed that in his context time showed a contradiction and he was right and logical to suggest that time did not exist, or is unreal. That was a sensible and economic view but slightly harder to develop than in Leibniz's case, where Leibniz had effectively inferred that space, per se, did not exist and was able to get quite a good theory for his era. But McTaggart's concern with time is in many ways very analogous with Buber's concern with space. Buber knew more or less what space was, but when he thought about it, it looked somehow spooky and unreal. Maybe we could say that that is "Angst". It is certainly a clear indicator that something needed to be done.. Anyway, the same thing happened to McTaggart with time, and as we will pointed out here, just as Einstein resolved Buber's philosophical worry about space, so too category theory can up to a point resolve McTaggart's problem with time. But that of course does not give us the right to ignore McTaggart's problem just as relativity has shown we should not certainly not have ignored Buber's problem. Just as in a way we have all been ersatz Leibnizians, prior to Einstein, let us importantly try to avoid continuing the same line of error with McTaggart, whether or not a resolution of his problems is more of a serious mathematical and philosophical challenge than Einstein's resolution of Newton's problem was.

Neurophenomenology and Category Theory

The term 'neurophenomenology ' was used by Varela, who also made a serious effort to understand consciousness (Varela 2000,2000a). It has to be said that it is a good idea to take his work at least almost as seriously as that of McTaggart (1927), and indeed Lawvere (2005).

In his day, till about 2001, Varela was probably at the forefront of neurophenomenology. However even as recently as the end of the last century, there was relatively little work on complexity theory and category theory as applied to neurophenomenology. The papers of the Ehresmanns (1999) gives an account of how the Ehresmanns at least, tried to use category theory. And references (Ehresmann, 1999) at least explain how it could be done. Some ongoing work is being done, for example, by Brown (2006). This indicates potential use of category theory which is anything but irrelevant and abstract (Note 1b)

Varela and the Specious Present

I should point out that though Varela wrote frequently about the specious present he does not seem to have ever actually used category theory as a working mathematical tool, nor to have given reasons why not. However both Varela and many others have clearly found coping with the specious present to be difficult, and certainly have not given convincing accounts on McTaggart's paradox. However when we read the account of Brown (2006), for example, we can readily work out that at least a meaningful account of the specious present can be made. At this point we are not unduly concerned with emergence, for which Brown thinks he may be able to obtain answers and, apparently, even mathematical results.

What we can certainly try to do now is to use colimits in a way like Brown et al (2006) tried to use them. There is a problem with Varela's work and it comes out clearly, for example, in section 3(2) ('The neurodynamics of temporal appearance') of reference (Varela, 2000a). I believe that one problem is that earlier workers have had to try to describe the McTaggart A series in terms of Newtonian time. Newtonian time is essentially punctal and in using it, we would have, very often, in effect to try to turn a blob into either one dot or into a series of dots. That is what happens to Varela. I will not give a bibliography here of all the other efforts to turn a blob into a dot, but they are common. For example, some of them are referred to in the references in Savitt (2002). Symbolic logic certainly produces some intricate formulae but those do not describe an 'instant in time' very well either.

Colimits and the Specious Present

We may not need to go quite so far as Brown et al (2006) does. We only need for the moment to consider an approach somewhat like that of the Ehresmanns. I append two diagrams from the Ehresmanns' (1999) study.

I will carry out this explanation in a way paralleling reference (Ehresmann, 1999), so that anyone who reads and understands (Ehresmann,1999) may be able to refer back to it directly to help to make it clear what I am saying here. There are many important differences to (Ehresmann,1999), however.

Now for anyone who has not a copy of McTaggart (1927) on hand, Professor Soshichi Uchii's rough one-page summary (Uchii, 2003), which does not go into all the subtleties of McTaggart's two volume book but will do for an introduction though is probably inadequate for the preparation or consideration of critical comment, is available on the internet. Uchii's summary at least tries to represent the A series as instants in time. (Don't worry about most of his comments or views at this stage).The B series can be a 'block universe' or some other punctal time representation that we care to use.

We don't worry about the McTaggart paradox as such at this point either, we just set up a McTaggart style representation.

We consider an instant P as a pattern of past present and future. This could be at this point the past present and future of the universe or of one object, say an observer, in a universe.

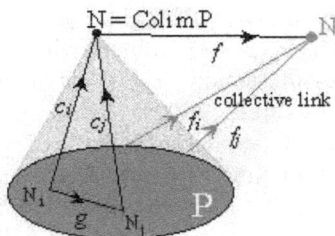

In a category, a pattern P is modelled by the data of a family of objects Ni and of some distinguished links between them. A collective link from the pattern to another object N' is a family of individual links fi from each Ni to N', correlated by the distinguished links of the pattern, in the sense that, if g is a link in P from Ni to Nj, we have gfj = fi .

The collective links model collective actions (constraints, energy, or information transfer) of all the Ni acting in cooperation along their distinguished links, and which could not be realized by the objects of the pattern acting individually. The cooperation can be temporary, as in a group of people who decide to cooperate for a particular work. But the association itself can be represented in the system by a more complex object N, which 'binds' the objects of the pattern and acts by itself as the whole pattern, in the sense that its links to any object N' are in 1-1 correspondence with the collective links from the pattern to N'.

In a category, the object binding the pattern (if it exists) is modelled by the colimit (or inductive limit) of the pattern. An object N is the colimit, or the binding, of the pattern P if it satisfied the two conditions:

1. there exists a collective link (ci) from the pattern to N,

2. each collective link (fi) from the pattern to any object N' binds into a unique

253

link f from N to N', so that fi = cif for each i.

If a pattern has a colimit, it is unique (up to an isomorphism). In this case, we also say that the pattern is a coherent assembly and that its colimit represents a higher order object which subsumes the activity of the assembly.

The colimit actualizes the potentiality of the objects to act together in a coherent manner by integrating the pattern in a higher unit (for example, the protein as such). In a natural system where the links have a given 'strength', the formation of a colimit is characterized in two ways:

1. 'locally and structurally', a strengthening of the distinguished links of the pattern restricts the degrees of freedom of the objects to ensure a more efficient cooperation among them;

2. 'universally and functionally', the actions of the colimit on the other objects of the system subsume the activity of the whole pattern (they correspond to its collective links).

For example, a molecule is the colimit of the pattern formed by its atoms with the chemical links defining its spatial configuration.

Roughly, the colimit forgets the precise organization of the pattern and records only its collective actions, and these can be the same for more or less differing patterns.

The rôle of the distinguished links of the pattern P is paramount: they determine the 'form' of the colimit and allow for the emergence of collective actions, transcending the individual actions of the objects. The coherence and the constraints introduced by these links can be measured by comparing the colimit to the simple amalgam of the objects Ni of the pattern, obtained if the links are forgotten, which is modeled by their sum.

Time Travel in Theory and Practice

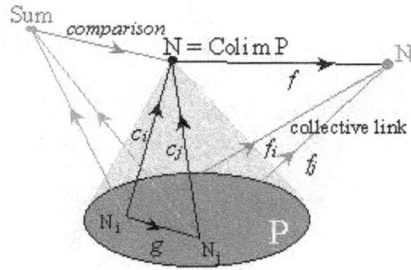

The sum (or coproduct) S of the family (Ni) is the colimit of the pattern P'
formed by these same objects but without any distinguished link. It classifies
the individual actions of the objects, while the colimit of the pattern P
classifies their collective actions made possible thanks to their distinguished
links in P. (Think of the difference between the behaviour of an unorganized
mob, and the behaviour its members adopt under the direction of leaders.)

There is a comparison link c from the sum S of the Ni to the colimit N of P,
which binds the canonical links from the Ni :to N. It measures the constraints
imposed to the objects by their distinguished links, hence by their participation
to a collective action. The links from S to an object N' which factor through c
correspond to the emergent properties of the complex object N compared to
the properties of its components Ni.

Now we could say that a series of 'instants' P, which we could call {P} could
occur as part of an ordered set or otherwise but we do not have to do this. And
each 'instant' has its own past- present-future. And a series of instants will exist
in some category Cp, say.

The specific 'instants' are not like a series of beads to be hung on a string, but
form significant but differing parts of a whole. In a sense each instant could be
taken as a past-present-future representation of some whole. The whole could
form a specific structure, possibly a category we might like to call MacA. We
bear this in mind as a structure, which like so many others, needs further
definition in due course dependent on circumstance.

But to sum up, I realised before too long that McTaggart's paradox, far from
being well understood - and indeed it now almost seems to some to be like an
effete toy for philosophers - still had not been resolved. There is in fact both a
McTaggart A series and a McTaggart B series, even if philosophers try to

pretend it has all been sorted out. We know that Zeno's paradox, for example, still has much to say, and so indeed has McTaggart's paradox. I will press on for the moment rather than to justify in detail. Philosophers still make a lot of money by discussing the pro's and con's of McTaggart's paradox, so I will not add further to the agony at the moment. (Chalmers, 2006)

So we have an A series and a B series, and we need to know what to do with them. Just an A series, just a B series, or two separate series which do not seem to map properly onto each other.

The A series in more detail

So what is really wanted now is something which gives solid physical prospects, such as more detailed dream experiments (as I tried to point out in Yates (2008)) or OBEs (out of body experiences) or NDEs (near death experiences). The mathematical and physical prospect of many worlds type interpretations is better than those of much of today's physics, in fact a useful quantum computer is likely to be built by 2020 (Ball, 2006), and this may help the process of such lines of understanding or intuitive interpretation. As these things go, the present supposed restrictions of any MWI is certainly likely to evolve in that time, and of course we have as yet no details as to how. But the A series for instance will probably turn out to be be a proper class (Note 1) and to begin with we may have to map a pseudo A series onto a mock B series to get results, and in effect I recently suggested something like that in my blog, Yates (2008). The 'block universes' of the B series type have been relatively easy to handle so far, though philosophically and to the intellect not altogether satisfactory, without an A series.. When we look at the A Series and the work of, for example, Varela and Ehresmann, but consider problems such as (and only "such as", certainly not exclusively) that the A series may be a proper class (that is, roughly, a class which is not a set) we are left with at least two obvious approaches, the first being to examine further immediately the A series properties of time and consciousness in terms of the work of Varela, Brown and Ehresmann for example.

Preliminary plans for a detailed MacA (or Category Theoretic Mctaggart A series)

Chalmers (2006) said : "The Time and Consciousness conference in Sydney yielded a lot of food for thought. The talks focused on a number of different connections between the phenomena .....There's obviously a lot of room for further work here, and I'm looking forward to seeing how things develop in coming years."

Chalmers is right about the conference providing a lot of food for thought and he is even more right in that there is a 'lot of room for further work'. In fact a lot of ideas but nothing solid yet (Note 1a). And I am hoping to provide something a bit more solid, as I have already tried to do in my blog, especially involving 'specious time" and category theory.

To be more explicit about the last paragraph: What I am finding from the conference details, and from earlier work, is that in problems being in and relating to the 'specious present', philosophers are frequently putting forward interesting discussions and concepts nowadays but that these on closer examination seem to have a circular or self-serving element. I choose as an example Kelly (2006) I quote "...the specious present, by nearly all accounts, lasts only a relatively limited time. Recent estimates generally agree that it is in the area of three seconds or so. But we often experience things to be moving for periods that are longer than this. If you watch an airplane taking off from the runway you can follow its continuous motion for several minutes before it disappears."

Great concept! Kelly then discusses the Retention Theory and its relation to perception experiments and philosophy. Well, for me the whole manifesto of such lucubration to date seems encapsulated in Alexander Pope's doggerel. "Remembrance and reflection - How allied ! What thin partitions sense from thought divide." In fact Kelly goes on by discussing Kant and Husserl and ends with "What we would like is a standard set of examples that give us the feel for what it is to experience something now as just-having-been." A good idea perhaps - but how? So I am left with the view that a more satisfactory category-theoretic interpretation has to be made and this will come closer to giving us a correct mathematics. When you think of it, perhaps he is looking for an extended A series here but may end up by conflating the lot with a B series, or something other writers may see as a B series.

In sum, a simple B-series interpretation of the world involving the physics of Galileo or Newton/Leibniz or Einstein is very adequate for some predictive descriptions. If we need to we further note that there is as yet no apparent compulsion (as in Le Provident (2006)) to consider Relativity (special or general) in any detail to start with, during our own continued lucubration. Hopefully it will not add to time-ordering problems in our A series or can be dealt with when it does. These problems, which Kelly and others mention, should either fall from our existing simple A series work, but very clearly manifest themselves in the A series discussions.

But there will only be relatively generalized answers at least to begin with, and this is not necessarily bad either. Consider the Baez example of a beautiful and possibly prehistoric use of category theory (Note 2). So we are trying to be like the shepherds of old, but not just inventing the B series as we could assume they did, but the A series!

Smith (2006) is probably also worth chewing on as the dynamic nature of time does seem to come out rather roughly in a scheme which proposes individuals existing at independent 'specious presents" i.e. a row of ...p0, p1, p2,..... etc. The idea of dynamic following is the hard one to include in the category and there is no reason why we are obliged to spell it out in terms of B series physics. However it exists as individuals exist in the frames ... p0, p1...etc. for each individual and the fact that we have mapped them on to a kind of ersatz (using Chalmer's (2006a) B-series word) A series or 'fallen' A-series does no harm.

The point is here that the MacA has to include dynamism whether or not some categorized or set theory version does. At the same time, present day mathematics has no simple format for providing dynamism within category theory statements or proofs. Certainly, the proof could be presented for example in the form of a video or even a notional mental head up display but this would not seem to present more actual mathematical content than the more normal pen and paper. Bearing in mind that at this point we are trying to present an A series in B series terms, this is not surprising.

And it is as well to remember that, say, the B series equation representing a ball rolling down an inclined plane does not need to be rolling down some inclined plane itself to be of immediate use. But this obvious fact is not the same as our current problem.

Indeed the obvious way is to try to write down a categorized version of MacA in terms of some categorized element pa where pa is a member of ...p0, p1,...is possibly to let pa be a present ism's p at time ta. This may help to eventually write the matter down in more detail in category theory terms. It sounds like a cumbersome multistage way to do things but may be appropriate. It should perhaps be pointed out that this process is not simply intended to result in an unnecessarily tautologous form of presentism but ultimately for enough positive description to allow an A series.

Time Travel in Theory and Practice

What to do then? From Chalmers (2006a) we could try to sort out the two-stage model (Notes 3, 4) and maybe relate it to Velmans's (2002) work - bearing in mind Chalmers claims to allow different models (presumably including Velmans's model with required justifications i.e. provisionally as I might do). The result may eventually be a new or a mutual model which could be multi-stage. However Note 4 is only an indication of possible proceedings; certainly the result should be expressible in terms of category theory.

On the face of it there clearly can be some form of mapping from 'real time' (or 'Eden time' or whatever it might be) to a B series time as such mappings have given many of our results in physics to date. To say that is almost tautology. Whether such a mapping presents a suitable or adequate representation or not is another matter. And certainly we may be now encumbered with an 'Eden time' A series and an 'apparently observed' A series. Whether this will be a simpler or better way to handle things, as yet we cannot be sure but one way would be to bear in mind a 2- stage representation as a possibility until the matter is more specifically concreted. I think we will need the A series even with the two stage model in any event. It could be that a 'de-Edenised' A series will not look very different to a 'de-Edenised' B series. The process which I would have in mind would be similar to or perhaps equivalent to obtaining categorification by first decategorifying and the categorification again, really not too different to what could normally be used in category theory anyway to obtain suitable categories from naive mathematical results in a fairly transparent way. (Note 5)

So I get to a point that a reasonable physical assumption seems to be that the A series is a Proper Class. Bays (2001), in "Reflections on Skolem's Paradox " says "if we start with a proper class which "satisfies" some finite collection of sentences, then the Skolem hull construction lets us find a countable set which satisfies the same collection of sentences".

Briefly what I am saying, and obviously there are many provisos, hedges and restatements, is that I consider a form of A series and a form of B series - but the A series is probably a proper class (like the somewhat similar class of all automata for example is a proper class (Adamek et al, 2004)) and probably cannot be effectively mapped, certainly not one to one, onto the B series. We only need to look at say Goldblatt (1984). In other words time is a rather complicated entity and when we get down to the mathematics or even the logic of time, we are using at least two different and not one to one mappable onto each other (mathematical) categories, A and B say (Note 4). And McTaggart had thus considered that he had found unreality in time when he was actually trying to compare two different things - though it is possibly not necessary to follow through his precise reasoning here, particularly on the C and D series. Clearly when considering a complex entity like 'time', there may ultimately be many more matters to consider but that does not remove the fact that McTaggart had found at least two such entities, the A and the B series.

In other words, I can use conventional complexity theory mathematics to study the A series as long as I remember that I am no longer in "block time" or B Series time. Now authors like John M. Gottman (2002) have used simple enough mathematics for years to solve psychological problems and seemingly eventually tried to shoehorn their ideas rather without thinking into a "block time" scenario where they will not properly work. And the heartbreaking discussions amongst proponents of tensed and non-tensed time may never again have to carry so much weight, (at the Sydney conference, (2006) they mainly did carry weight) in a situation where a tensed time (A series) is used where convenient for people, and a non-tensed time (B series) where convenient for objects. And the maths can be great in both cases. Though of course that does not say it will be straightforward or easy.

Also the fact that we are using not a real A series but a pseudo A series mapped into the B series, undoubtedly changes the mode of operation and the character of any work we may do. And in sum, the above tends to dispose of the somewhat short sighted idea, which I feel may have been despised by such as Emmy Noether, that we only need deal with the B series anyway. Then we would not have a past, present and future, but only idealised 'block time', which is in no way a complete description of the universe though it has served physics in a limited fashion for generations.

The Many Bubble Interpretation

The 'Many Bubble Interpretation" or MBI, (Yates, 2008a) appears by means of a model of the McTaggart A series. Without being initially sidetracked into the fascinating coherentist theories of epistemic justification, we simply loosely define A series bubbles for present purposes as being entities inside which a person, persons or whatever are for the moment severally confined, each at some personal present (which we know from as far back as the work of Kleinhuber, Libet, etc., is not readily defined as a single point in time, but more usually is taken by psychologists and others to have at least some ongoing 'duration'), and with a past, a present and a future, in accord with the spirit of the McTaggart A series. The work of LePoidevin, Quentin Smith, Dean Zimmerman and many others is borne in mind. And as Dyke has said, we may not be forced to countenance plurality of further worlds in such circumstances – although we can. The A series is treated as a large category, intrinsically unmappable one to one onto the B series. There is also a B series and this can often be represented by a quantum mechanical description of the universe.

So we have both an A series and a B series, and McTaggart's work and Zeno's work, (and/or their modern counterparts), can pose no problems. Now in practice, since the A series will almost certainly be a proper class, we still have not pinned it down in great detail and indeed may because of its nature, never to be able to do, so we are using a pseudo A series written in the B series. Obviously there will be lacunae and these lacunae may exhibit themselves partly in the form of the quite hard to describe   equations we will be immediately confronted with.

Outline of the MBI

A relatively simple basic mathematical model for a "bubble" in the MBI ('Many Bubble Interpretation') discussed earlier, can be constructed. For a 'look and feel' description, see Note 5. Many bubbles – and there would be many – could be much the same, in principle, and given by , for example. And in the simpler cases of the model there need not exist episodic memories to retain many of the apparently intrinsic features of human thought (Egan, 2007). Even total loss of personal memory made no difference in subjects tested. Indeed Rosenbaum (2007) goes so far as to say: "We found that if you're trying to put yourself mentally in someone else's shoes, you don't need to put yourself in your own shoes first." We do not even need, of necessity, to consider mirror neurons to 'have a life'. We can even, in terms of level of simulation simplification, try to emulate Winfree. And no complex 'Theory of Mind' is required (Ramsay, 2007).

There is no need to deal at this juncture with the problems posed by Honderich or by, for example Trevena and others, to the work of Libet (2003), and its defence by Haggard, Klein and others. Libet's results, or others, will just be part of the Madonna formalism within the bubble, which can be "pseudo A series" in its formulation, I think.

Obviously, more complex contents can be given to the MBI and this is being done.

Applications of the MBI (example only)

A concrete case of the application of the application of the Many Bubble Interpretation appears in Complex System Theory. I used the standard Dynamical Systems methods as described in Hannon and Ruth (1997). Whilst that book recommends the use of the program Stella, I in fact used Berkeley Madonna (8.3.11) but the results come out in a somewhat similar format. In fact I used a Romeo and Juliet type model (Sprott, 2004) (Yates, 2008b).

There are many potential uses of such a model, and various examples have been considered. A very striking example, though one out of many, is the example of what may happen when we dream. This seems much more flexible than many other cases - synaesthesia, for example, is a difficult though potentially rewarding one, and this as well as other cases is being documented in the literature. (Yates, 2008, 2008c)

Clearly one direct entry to a study of the unconscious could be by studying dreams, without any necessary preconception as to what these dreams are, or what they mean - if anything. A brute fact about dreams is that they exist, and can thus be studied. Studying dreams is what the shaman did (Charlton, 2002), and we can do the same. As Charlton (2003) pointed out "The altered states of dreaming consciousness enables hunter-gatherers to cross further boundaries of time and space in pursuit of high-level insights that synthesise and integrate complex knowledge of many kinds ", but he then goes on to say that using systems theory would make this difficult. In fact, to date most studies of dreams look relatively speaking rather basic, to those who want an early quick explanation of such matters. There are interesting exceptions and I mention the experiments of Stickgold, and the Stickgold effect (also perhaps better known in the vernacular as "Tetris Dreams", because of the frequency it was noticed anecdotally, and "Tetris Dreams" even gave over 1500 Google results including a T shirt, but of course the effect can be located in many other cases than simply playing Tetris, and it has interesting neurological implications also).The Stickgold effect (1999, 2000) is a pretty simple idea, in essence - one performs a simple task like playing a computer game, and then dreams about it. So apparently the Stickgold effect induces dreams. We found that there also seems to be a reverse Stickgold effect, where the dreams are dreamt and subsequently the acts dreamt are performed, within the confines of a controlled experiment using double blinds, for example the subjects are kept in the dark about the experimental details during the experiment.

Now there are real problems in assuring reproducibility of results and I for one

Time Travel in Theory and Practice

will not be happy until at least we reach the high levels of experimental reproducibility obtained in the early Milgram (1974) experiments in general psychology. Preliminary experiments are proceeding at our Vasai, India address at the moment but we are also considering setting up a Science 2.0 style site which will allow the comparison of varied psychological experiments, some of which may contain physical data like DT-MRI data as well as the normal psychological profiling data. Such sites are now relatively common in genome experiments and there seems no reason why the idea cannot be usefully extended to many psychological experiments such as those on the Stickgold effect and the possibly newly discovered reverse Stickgold effect. Perhaps modified versions of the Science 2.0 idea can be described as Psychology 2.0 or in the case where applications could be said to 'transcend' or supplement science, and experimental philosophy results are in the offing. then it can be described as Philosophy 2.0., where we might wish to consider such workers as Chalmers and Knobe (2008). In the present context of the Many Bubble Interpretation there are already possible equations of constraint obtained using the modelling tool and the simplified form of the equations in the present mode, described in more detail on the website of Yates (2008b), is

$dR/dt = a*R + b*J*(1-|J|)+e*Z$
$dJ/dt = c*R*(1-|R|)+d*J +f*Z$
$dZ/dt = h*S +g*R$ (S is Heaviside step functions :in N003a,g=f=0)

The next step may we involve the refinement or replacement of the present equations in Berkeley Madonna using methods of Self Organised Criticality, and in particular the use or incorporation of a mode like the sandpile mode may help. If these equations can be improved and/or accurate limits set on their parameters, they could be used for yet more tests and even more accurate results, in for example the mode, duration and timing of stimuli. We bear in mind Winfree's work as a parallel example of such methods. In the above equations, very roughly, R ('Romeo') and J ('Juliet') effectively represent the 'unconscious' and 'conscious' mind or equivalent representations in other philosophical approaches,and Z the applied impulse. The notation is like that of Sprott (2004). The improvement in the equations will likely be carried out along with attempts at semi- empirical assessment of the physical factors under consideration.

In Quantum Theory:

After a consideration of the MBI, the Schrodinger Cat Paradox ceases to seem like a paradox and in a poster (Yates, 2008d) I illustrate this and further

263

examples of the simplification of an understanding of quantum theory and related topics like quantum computing, even including Kwiat's work. The Schrodinger Cat description is in Note 6, equations are as in Yates (2008b) plus conventional quantum theory..

A Specific Application:

A problem involving some applied mathematics and philosophy.

The Many Bubble Effect described herein, together with other factors like McTaggart's paradox and Zeno's paradox, allowed a formulation in terms of differential equations of Stickgold's dream experiments and my interpretation and furthering of them. This led to a number of equations and graphical results. In particular to equations like that described as N003b on my website at http://ttjohn.blogspot.com/ (RSS feed available) and on the CD.

Very briefly, as the 'pseudo A series' might describe it, there could be tiny pushes and impulses to the mind at a given time, from both past and future stimulations, but at a particular time it could be said that the mind is in some kind of dynamic balance which Stickgold altered in the 'Tetris dream' by a push from the past, relatively easy in retrospect. In my case I alter the position of the push from the future to the present, and this worked too. Experiments and trials are still under way, and could show conclusively the merits of the MBI, though their success is not essential to it. And a sandpile Madonna model is considered in Note 7.

Notes

Note 1. Goldblatt's 'Topoi' refers to a 'proper class' as 'a class which is not a set'.

Note 1a. I am very sympathetic to those who succumb to this problem. For example, a point was made very clearly on Yogacara Network (http://www.yogacara.net/node/2443) "How much pomp, circumstance and apparatus academia requires in order to frame even a very small and simple point. References to everything in the literature ever said on any vaguely related topic, detailed comparisons of your work to whatever it is the average journal referee is likely to find important -- blah, blah, blah, blah, blah.... A point that I would more naturally get across in five pages of clear and simple text winds up being a thirty page paper!" It is not going much further to retreat from substance to form and to provide only self-serving form.

Time Travel in Theory and Practice

Note 1b. On the 'abstract nonsense' issue, a term which is unfortunately sometimes used about category theory. Category theory is anything but 'abstract nonsense', a term which may have sounded funny as a mild joke for the cognescenti, many years ago. The term could actually be very misleading today for an important but difficult topic which is certainly not very abstract anyway. I have seen genuine confusion on the internet in cases when people in some way innocently ascribed a literal meaning, even a wouldbe positive meaning, to 'abstract nonsense' as a term for category theory. The term 'abstract nonsense' was apparently devised as an 'in joke' by Norman Steenrod, a topologist, (1910-1971), who in fact liked category theory and found it immensely useful. But Birkhoff and MacLane's "Algebra" or Geroch's "Mathematical Physics" as 'abstract nonsense' ? Neither book is particularly abstract and certainly not nonsense, even metaphorically. And both books are brim full of the practical uses of category theory! Further, computer people seem to have found many non-abstract uses for the subject to the point where the term 'category theory' has almost been hijacked by the computer people, so that people almost think it is a computer technology term not a mathematics or fundamental physics term. (Historically the readable "An Application of Abstract Nonsense to Surface Area" by Harriet Lord, illustrates the use of category theory).

Note 2. arXiv: math. QA/ 9802029 v1 5 Feb 1998 Categorification. John C. Baez. This says: "Long ago, when shepherds wanted to see if two herds of sheep were isomorphic, they would look for an explicit isomorphism. In other words, they would line up both herds and try to match each sheep in one herd with a sheep in the other. But one day, along came a shepherd who invented decategorication. She realized one could take each herd and `count' it, setting up an isomorphism between it and some set of `numbers', which were nonsense words like `one, two, three, . . . ' specially designed for this purpose. By comparing the resulting numbers, she could show that two herds were isomorphic without explicitly establishing an isomorphism! In short, by decategorifying the category of finite sets, the set of natural numbers was invented."

Note 3. Chalmers (2006a) states: "It is a further question how this model should be extended to the representation of time and motion. I am inclined to say that the two-stage model can be extended to time as well as to space, though this turns on subtle issues about the metaphysics of time. A natural suggestion is that the Edenic content of temporal experience requires A-theoretic time, with some sort of true flow or passage. Our own universe may

265

not instantiate these perfect temporal properties, but it may nevertheless instantiate matching B-theoretic properties (involving relative location in a four-dimensional "block universe") that are sufficient to make our temporal experiences imperfectly veridical, if not perfectly veridical. The representation of motion could be treated in a similar way." and "One might go so far as to suggest that Eden is a world with classical Euclidean space, and an independent dimension of time, in which there is true passage and true change. Our own world is non-Euclidean, with time and space interdependent, and with pale shadows of perfect passage and change. On this view, Einstein's theory of spacetime was one more bite from the Tree of Science, and one more step in our fall from Eden."

Note 4. Chalmers (2006a) on two-stage: "the two-stage view yields natural answers to the objections to the Fregean view that were grounded in phenomenological adequacy. On the relationality objection: the two-stage view accommodates relationality by noting that there are certain specific and determinate properties—the perfect color properties—that are presented in virtue of the phenomenology of color experience.

When Jack and Jill both have phenomenally green experiences in different environments, the two experiences have a common Edenic content, and so both are presented with perfect greenness. This captures the intuitive sense in which objects look to be the same to both Jack and Jill; at the same time, the level of ordinary Fregean and Russellian content captures the intuitive sense in which objects look to be different to both Jack and Jill. By acknowledging Edenic phenomenal content in addition to Fregean phenomenal content, we capture the sense in which perceptual phenomenology seems to be Russellian and relational."

Note 5. In my mind, I tend to think of the A series as being like a lot of bubbles floating freely, each of which representing a person or sentient object, and his or her past, present and future at some time, and we could hopefully index the persons in the bubbles as (Pn,Tm), this being person Pn at time Tm. By now the apparition has degenerated to a pseudo A series (almost nearer to being a B series). But in principle we are mapping an A series to some model we can understand. And if we want we can follow memes through the bubbles by now, and index like (Pn,Tm,Mi) where Mi is some meme which may occur as part of one or more bubble. But this is intended as a guide rather than mathematics or metaphysics.

And whilst as presented above, the A series has a "future" along with each "present" and "past", in the individual bubbles. This is only a model and not a metaphysical description of the universe. It is however by the nature of the model, many world in structure. The claim is not made that these many worlds have to exist in actual fact. So the MBI ("Many Bubble Interpretation") seems to be in basic distinction from the MWI of Price or Deutsch or indeed the MCI ("Many Computations Interpretation") of Mallah (2007). The latter two are in origin B series, and to aid consistency should possibly be assumed to exist, in some sense, at least in the sense that the quantum mechanical results in Hilbert spaces exist. In the MBI, the many bubbles of time, each with its own past, present and future, are as real as the person or conglomerate observing them, and only exist in a model of the A series. The A series itself, in some metaphysical sense at least, can be taken to exist. So the Baldwin (Note 5b) style bubble referred to above, will contain the person (TRB) at the indexed time in the bubble in York, with a past somewhere else, perhaps partly in Leeds, and presumably a future somewhere else again, perhaps partly in Blackpool. This will simply be at the 'time' referred to above for TRB, and the bubble is only part of a model which contains many more bubbles. But this is only part of a model of the A series.

And, without even invoking quantum theory, Quentin Smith (2002) explains how some models of the A series can seem to have B series concomitants, even in special relativity. In fact if we wish, we could consider our A series bubbles corresponding to different Tm values to be linked to one another by a spider web of gossamer chains. The spider web could now seem to be very clearly savouring of the B series, although we had started with a model based on the A series. Given suitable provisos, that spider web might well suit STR (Special Theory of Relativity).

Furthermore, the Schrodinger Cat riddle seems to give no essential problems in the MBI, and the MBI has the additional virtue of flagging up the obvious apparent anomalies that the Cat paradox has seemed to show to some, in the B series (Note 6 gives more details).

Note 5a. "This point connects with a deep distinction between practical and theoretical points of view. The practical point of view is essentially 'first-person' ('subjective'): it assumes knowledge of who I am (TRB), where I am (York), what time it is (today's date). The theoretical point of view is essentially impersonal ('objective'). It doesn't require this first-person knowledge. So the A-series/B-series distinction is a case of the distinction between these two points of view, the practical and the theoretical."

Note 6. In the case of the Schrodinger Cat Paradox, as shown on the poster (Yates, 2008), instead of having the eponymous Cat in a room with a bomb and a puzzled query in the mind of persons outside the box, this neatly splits into roughly into three cases, in the present treatment.

(1) A series: Cat and observer, each in their own bubble, no way that we know of so far that the 'observer' bubble can get at the 'cat' bubble. So no paradox.

(2) Pseudo A series: We might try to simulate the cat ideally in the observer bubble, for example. Category theory suggests how. We try procedures as in "Applications of the MBI" in the main text above.. N.B. May use 'B' series maths.

(3) B series: cat and observer have the same math structure so far. Not a complete description but mathematically OK.

Note 7. We can now consider further simple ideas like using the sandpile analogy as it has been tried, for example, with software development, without the physics actually disappearing from the system as the actual software used for development does in the paper (Wu, 2002). That contained an excellent analogy:

Driving force / sand drop /change request
Response / sand slide / change propagation
System state / gradient profile / release, iteration plan
Relaxing force / gravity / stakeholder satisfaction

but plainly "the mathematics was not the territory" just as "the map is not the territory" to a geologist. Even if a geologist goes along with the mathematical fractals approach, it does not get the dirt under his fingertips. But with the MBI approach in physics, we seem to be as close to a physical simulation of the real world as we can be at the moment. Importantly, for example, we have not simply beaten off McTaggart's paradox but on the contrary, we we have used it as strongly as we can.

This may well give us the ability to prepare a more precise or even a new and better Madonna model than the model N003b suggested in an earlier entry, using for instance some of the methods of Dhar (2006) particularly as described in Dhar's section 3 onwards and other such SOC methods, as well as what we are using to date.

Acknowledgements

I would like to thank Deepak Dhar and Navin Singhi, both of the Tata Institute of Fundamental Research, Mumbai for their helpful advice, encouragement and discussions, and Deepak Dhar for the chance of having a fine lunch at the Tata Institute of Fundamental Research, Mumbai. I would also like to thank David Chalmers, Professor of Philosophy at the Australian National University, for his helpful advice and encouragement .

References

Adamek J., Herrlich H., Strecker G.E. (2004), "The Joy of Cats", p15, John Wiley , and katmat.math.uni-bremen.de/acc/acc.pdf

Buber, M., (1959) Between Man and Man [1947], trans. Ronald Gregor Smith, Beacon Press, Boston.

Ball P., (2006), 'Oxford University reckon that "A useful (quantum) computer by 2020 is realistic,"' Nature 440, 398-401

Bays T., (2001), Ph.D. Dissertation. "Reflections on Skolem's Paradox" p86, UCLA Philosophy Department.

Brown R., (2006). For example Brown R., Paton R., Porter T., "Categorical language and hierarchical models for cell systems", to appear in Computing in Cells and Tissues - Perspectives and Tools of Thought', Springer Series on Natural Computing. More preprints of material like (Brown, 2006) such as 05.13 are on Brown's home page at http://www.bangor.ac.uk/~mas010/ or at http://www.informatics.bangor.ac.uk/public/mathematics/research/preprints/

Chalmers, D. (2006)
http://fragments.consc.net/djc/2006/07/time_and_consci.html

Chalmers, D. (2006a) "Perception and the Fall from Eden" (T. Gendler & J. Hawthorne, (eds) Perceptual Experience. Oxford University Press, 2006.)

Charlton B.G., (2002) "Alienation, Neo-shamanism and Recovered Animism"
http://hedweb.com/bgcharlton/animism.html

Charlton B.G, Andras P. , (2003), "What Is Management and What Do

Managers Do? A Systems Theory Account" , Philosophy of Management , Vol 3, p3-15

Dhar D, (2006), "Exactly Solved Models of Self-Organized Criticality", http://theory.tifr.res.in/~ddhar/leuven.pdf

Egan L. C., Santos L.R., Bloom P. (2007), "The Origins of Cognitive Dissonance: Evidence from Children and Monkeys". Psychological Science, 18, 978-983.

Ehresmann A., and Vanbremeersch, j-p. (1999) "Memory Evolutive Systems", http://cogprints.org/921/ ; also see Ehresmann C., http://perso.wanadoo.fr/vbm- ehr/Ang/W24A5.htm , http://perso.wanadoo.fr/vbm-ehr/Ang/W208.htm

Feynman R. P., (1982), "Simulating physics with computers", International Journal of Theoretical Physics, 21:467-488. Goldblatt R, (2008), "Topoi", p10, Dover

Gottman J.M., et al., (2002), "The Mathematics of Marriage", MIT Press , ISBN : 0-262-57230-3

Hannon, B. and M. Ruth. (1997) Modeling Dynamic Biological Systems. Springer-Verlag, New York City, New York.

Kelly S.D. (2006) The Puzzle of Temporal Experience Sean D. Kelly Princeton University To appear in Philosophy and Neuroscience, eds. Andy Brook and Kathleen Akins (Cambridge).

Knobe, J, (2008), "What is Experimental Philosophy?" The Philosophers' Magazine, (Forthcoming). Viewable at http://www.unc.edu/~knobe/ExperimentalPhilosophy.pdf.

Lawvere C.W., Schanuel S.H., (2005) "Conceptual Mathematics", Cambridge. and see Note 4.

LePoidevin, R. (2006) "The Experience and Perception of Time", The Stanford Encyclopedia of Philosophy (Winter 2004 Edition), Edward N. Zalta (ed.), URL =. states: "I ignore here the complications introduced by the Special Theory of Relativity, since tenseless theory—and perhaps tensed theory also— can be reformulated in terms which are compatible with the Special Theory."

Time Travel in Theory and Practice

Well the matter can be argued either way but it is fair to say, with LePoidevin, that special relativity is probably easiest left out of it provided we tie a proverbial piece of string to our finger to remind us of it if actually need be.

Libet B.,(2003). "Can Conscious Experience affect brain Activity?", Journal of Consciousness Studies 10, nr. 12, pp 24-28 ; and many others.

McTaggart J M E, (1927), "The Nature of Existence" , Cambridge: Cambridge University Press.

Mallah J.,(2007), arXiv:0709.0544 "The Many Computations Interpretation (MCI) of Quantum Mechanics".

Maxwell N., (2004), "The Ontology of Spacetime", Conference in Montreal, 14 May 2004

Milgram, S. (1974) "Obedience to Authority", Harper & Row, USA

Ramsay T.Z. , (2007), Science and Consciousness Review, November 26, 2007

Rosenbaum R.S.,Stuss D.T.,Levine B., Tulving E.,(2007),"Theory of Mind Is Independent of Episodic Memory", Science, 23 November 2007: Vol.318. no.5854, p. 1257 DOI: 10.1126/science.1148763

Savitt S., (2002) "Being and Becoming in Modern Physics", The Stanford Encyclopedia of Philosophy (Spring 2002 Edition), Edward N. Zalta (ed.).

Smith, Q. (2006) http://www.qsmithwmu.com/reference_to_the_past_and_future.htm , Time, Tense and Reference (eds. A. Jokic and Q. Smith). M.I.T. Press, forthcoming

Smith Q., (2002), "The incompatibility of STR and the tensed theory of Time", Published In: "The Importance of Time", editor, L. Nathan Oaklander. Kluwer: Philosophical Studies Series.

Smullyan R., (1961), "Annals of Mathematics Studies", Study 47, "Theory of Formal Systems", especially Chapter 3, Princeton University Press, L.C. Card 60-14063

Sprott J.C., (2004), "Dynamical Models of Love", Nonlinear Dynamics, Psychology, and Life Sciences, Vol. 8, No. 3, July, 303-314 ; and many others

271

including Aks, D.J., Sprott, J. C. (2003), "Resolving perceptual ambiguity in the Necker Cube: A dynamical systems approach", Journal of Non-linear Dynamics in Psychology & the Life Sciences, 7(2) 159-178.

Stickgold, R., Malia, A. & Hobson, J.A. (1999) "Sleep onset memory reprocessing and Tetris. Journal of Cognitive Neuroscience" 11(supplement)

Stickgold, R., et al , (2000), "Replaying the Game: Hypnagogic Images in Normals and Amnesics" Science 290 (5490), 350. [DOI: 10.1126/science.290.5490.350]

Sydney conference (2006), "Time and consciousness Conference", http://fragments.consc.net/djc/2006/07/time_and_consci.html

Uchii S., (2003) http://www.bun.kyoto-u.ac.jp/~suchii/mctaggart.html

Varela F.J. , (2000) "Naturalizing Phenomenology: Issues in Contemporary Phenomenology and Cognitive Science" Edited by J. Petitot, F. J. Varela, B. Pachoud and J-M. Roy, Stanford University Press, Stanford Chapter 9, pp.266-329

Varela F.J. , (2000a) "The Specious Present : A Neurophenomenology of Time Consciousness Francisco" in: J.Petitot, F.J.Varela, J.-M. Roy, B.Pachoud "Naturalizing Phenomenology: Issues in Contemporary Phenomenology and Cognitive Science", Stanford University Press, Stanford

Velmans M., (2002) "How could Conscious Experience affect Brains ?", JCS, 9 (11), Special Issue, Imprint Academic, UK & USA, ISBN 0907845 39 8

Wu J., Holt R., (2002) "Seeking Empirical Evidence for Self-Organized Criticality in Open Source Software Evolution", Available Research Reports, David R. Cheriton School of Computer Science, University of Waterloo, Ontario, Canada

Yates J., (2008) ,"TSC 2008", April 29, 2008, and throughout site, http://ttjohn.blogspot.com/ ; and "Towards a Science of Consciousness", p147-8, April 8-12, 2008, Tucson Convention Centre, Tucson, Arizona, Center for Consciousness Studies, University of Arizona.

Yates J., (2008a), work to be archived or published elsewhere and in part in ttjohn.blogspot.com ; in fact the Many Bubble Interpretation, proofs for the importance of its use, and current applications, are to be discussed in more detail in separate papers.

Yates J., (2008b) , ttjohn.blogspot.com , especially including "Work in Progress on application of dynamic systems theory to the A series (1) and (2)" at http://ttjohn.blogspot.com/search?q=dR%2Fdt&x=45&y=8 etc

Yates J., (2008c) , http://ttjohn.blogspot.com/2006/07/precognition-dreams- and- mctaggarts.html#links "Can dreams predict the future ? : Experiments and considerations of them". (The answer to the question is "It is not easy but I am working on it" as per the top paper here).
http://ttjohn.blogspot.com/2006/04/do-we-dream-of-future.html and elsewhere in this blog. This is not a simple 'precognitive' effect and that is not claimed. Point 4, "Self-Organised Criticality - a possible tool for the MBI " on January 26th, 2008 for example, illustrates this point.

Yates J., (2008d) , Poster at http://ttjohn.blogspot.com/2008/04/tsc-2008.html and elsewhere in blog

7.09 Neurophenomenology and Category Theory

Clearly most people believe that the human brain is still active during sleep states and I now consider sleep states.

Current fairly acceptable views include those of Domhoff and Hobson. Metzinger has also widely expounded his views. In the first section I consider mainly Domhoff's views insofar as they relate to the present work, in the second section mainly some of Hobson's views (and mention fascinating experiments to confirm the applicability of the A series), and in the third section I consider some of the views of Metzinger and others. I give yet more possibilities, including those in lucid dreaming

Section I

Domhoff (2002) uses the Hall/Van de Castle system of dream content analysis. This is a reasonable approach. On his site he does not give much detail on case histories and he only gives one case ('Jeff') of lucid dreaming. We know or believe that there are many other cases of lucid dreaming though interest in the subject is perhaps currently not as great as it has been.. This suggests that for our purposes a slightly more detailed appreciation of cases and circumstances may be required.

Our point here is not that we wish to resuscitate the ideas of Freud or Jung although as usual we keep all historical ideas in mind, but our point is that since past present and future all need to be considered, not necessarily along a sort of one-dimensional historical path, we must bear in mind the past, the present, and the future of dreams as well. We must also bear in mind their conceptual relevance at past and future times. There is surely nothing irrational or bizarre about this. Fortunately Domhoff's dream content analysis at least contains the principles of how this can be done. But for case histories containing future events we are rather short of details so far. The 'body/mind/spirit' websites contain some examples but the study of statistics by for example Susan Blackmore (2006) and many others suggests that many of these results unduly emphasise favourable predictions. [It is essential not to hold out any workable theory as a hostage to fortune because of the unduly optimistic ideas of statistically impaired enthusiasts!]

We are not trying to predict the future, merely to present an acceptable and scientifically viable explanation of the world as it is. If we consider human beings, taken individually clearly each has an autobiographical memory however complete or inaccurate, a present time and a future time. The present time is roughly defined well enough although there sometimes seem to be lacunae, for example Libet's 'missing half-second', the doubts about free will given by Wegner (2002) and others, and so on. The future time is something which remains in the hands of such ideas as those of Bierman (2002), and philosophers. For the moment, Bierman's interpretations remain unconvincing to the general scientific community as do Hameroff's. As for me, I have many times stated that we have no good evidence whatsoever yet of a quantum brain effect following the Penrose model. Bierman's comments have yet to be proved and they may well be. Philosophers, on the other hand, reasonably place the future at our disposal. With little doubt, for example, the sun will rise tomorrow and Newtonian physics will, within stated limits imposed by relativity and other factors, hold for the indefinite future. It does seem as if, from the standpoint of individuals, past time remains in our autobiographical memory, present time is with us and future time is 'philosophically' with us, perhaps in less practical detail and with less certainty than past or present. The difference between past and future could be taken as to some extent quantitative rather than qualitative and we are all 'sure' of events in the past which in fact never happened (Newhousenews, 2006). We are just as 'sure', or even 'surer' of the sun rising tomorrow or of the Laws of Nature being unchanging at least till next week or of the Second Coming of Christ (if we believe the Christian Book of Revelations). Is there a difference between such certainties ?

Specifically, in certainties relating to past and certainties relating to future? With regard to the Second Coming, perhaps we can place that in a 'belief' bracket along with 'Noah's Ark' (this being a highly selective and personal belief bracket) but tomorrow's sunrise looks about as close to a certainty as yesterday's sunrise (for most people in practice). In fact, as I pointed out about 25 years ago, yesterday's sunrise is far from certain. That implies that maybe we all suffer from a form of universal anosognosia or are subject to some odd mental effects. This is a more likely philosophy to most people than solipsism is. Indeed, as David Deutsch (1997) pointed out as a joke (that way it sounds a little shallow), a solipsist philosopher might sometimes say at a gathering: "It's good to see so many solipsists together".

Those matters are far from simply metaphysical. We do not want to need to go 'metaphysically' (and we recall that Karl Popper refers to metaphysics as 'nonsense' - perhaps he had an extreme attack of logical positivism?) to the limits defined, for example, by Bostrom (2006). But, just as we saw in an earlier post, in view of mesoscopic results we cannot live forever with Copenhagen, so too we cannot live with a very simple mental picture of time and reality. And we need to look and examine time and reality. But it is probably not a good idea to assume that the new answers must necessarily come from quantum physics.

However one step that is immediately available is to apply a form of concept analysis using the Hall/Van de Castle system and try to establish past and future correlations between the dream analysis results and the appropriate case histories. This inevitably is going to involve more difficulties than we might expect if done adequately, for various reasons, some of which I will try to explain later in this piece.

Section II

I will move to a consideration of Hobson's (2002) work.

Hobson (2002) makes the important observation (H5) (references as (H..) refer here to page numbers in Hobson (2002)) that percepts and feelings make an important contribution to dream content and this often disregards time place and person. So colour, size and dimension are often less important contributors. Maybe in this sense, abstract paintings are more 'real' than physical appearances. So it looks as though dream descriptions or features could come closer to an A series description than to a B series description. Indeed dreams almost look like what we would expect from a foray into time-travel in a MWI. It is thus not too surprising that dreams have often and indeed are often believed to have precognitive qualities. (H6) reasonably instances other cases like linguistics and poetry where form and content. are at the very least complementary. (H7-8) Here he mentions three types of dreams and this attempt at dream classification may be important. The report 3 (H10) refers to the bizarre phenomena which all dreamers encounter, more usually in REM (as opposed to) and he, and indeed I, feel that these phenomena seem to demand an explanation in some real terms. (H11) He also claims that some brain functions increase, and some decrease, during sleep. This seems unarguable and it is clearly advantageous to find out which ? and why? (H12) Hobson has 116 volumes of his own dreams and some of these must be of value in interpretations etc.

He also says (H17-18) that some dreams are 'without interpretation', presumably for example along the lines of Freud or Jung and I certainly accept this as a working hypothesis.

(H20-21). He states that dreams are not precognitive Up to a point his arguments hold water. He says that some 'precognitive' dreamers will score a few hits, for example, but many more possible so-called 'hits' will not work. This is probably true, and is an argument often advanced reasonably
against precognition and similar notions, but later in this essay I will show that the circumstance fits in even better with A series reasoning. This also applies exactly to a precognitive style dream Hobson (2005b) later discloses, as do his arguments and my own.. With the A series we are simply trying to describe the universe as it is, and hopefully to factor in enough elements to allow our theories to be disproved.. So we take his point about his views on precognition, and remember that they amount to the view of a sincere experienced man who has recorded more dreams than most. But that in itself does not disprove 'precognition'. On balance, therefore, he has not discovered precognition and does not expect to. Logic suggests that either there is an ultimately verifiable reason for this or that precognition never occurs. So we are left with a new phenomenon (strange 'hits') and an explanation. The explanation is that they are 'only chance hits' and 'overall they average out'. I doubt if there is any really satisfying statistical proof for this statistical

averaging but it sounds fairly plausible without good reason to the contrary, and any advanced statistics student could presumably do a fair justification of his idea, as distinct from a good proof. Bearing in mind Occam, that should be enough unless there are other answers. I will have another possible explanation for his phenomenon.. Fortunately it will not disprove the A series (which has been solidly argued for, for many years) if it is wrong and it may not even help us with equities, cards or horse racing, but it could eventually prove that the A series is a worthwhile concept . We can at least hope for that, no mean feat.

(H22-23). At this point Hobson genuinely tries to compare the dream state with waking psychosis, an interesting analogy and possibly even creating a potentially more revealing and less dangerous topic of research than some induced brain stimuli such as the valuable rTMS. There are, of course, many differences from approaches like rTMS. One obvious difference (for say rTMS) is that both induced changes in rTMS are normally more specifically directed to particular parts of the brain and, within limits, they are more immediately controllable for specific ends. Another difference is that many people clearly dream for a lifetime without any obvious ill effects, and this would be unlikely for any invasional treatment I can think of carried out over a lifetime of patient study, at least at the present state of the art.. At any rate (H23) applies a 'mental state examination' and it surprises him that it has not been done before. One interesting fact that we may note already is that (H30) dreaming is hyperassociative. The nervous system also appears (after Helmholz) to create its own image of the expected consequences of movement.

On (H31) we note that he lectures the Freudian school for trying to make too much from hyperassociations but then promptly proceeds to do the same thing himself on (H33) by attempting to suggest a 'brain/mind isomorphism.'. The meaning of 'isomorphism' (or for that matter 'mind') is apparently not clearly defined at this point and seems to have as yet no accurately defined practical mathematical meaning here. But bearing in mind for example the great hopes, unresolved for 50 years, that computer specialists and linguists have had for machine translation, this is less of a criticism than a simple clarification of the fact that we are still working in as yet uncharted territory, and it is good that he seems to realise this by implication.

He says that his 'isomorphism' means that just as there is an awakening of the mind during sleep there is a similar though not identical awakening of the brain during sleep. (H34) particularly underlines by examples the possible likely benefits of his present approach. He obviously sees that there are problems as by (H55) he refers to the 'hard problem' and (H61) is bringing out some key questions as to the differences in brain activation during sleep. These differences during sleep are all important to any study which utilises sleep.

There are further important implications (H68) to any experiments involving sleep in that he seems to be able to actually initiate dreams and to enhance REM sleep by chemicals.

(H75-76) He claims to be always at the centre of dreams but surely he must be in some way conscious to know that a dream is his own. That is, for all he knows the mind is having completely 'unconscious' dreams (or brain patterns tantamount to dreaming) where he is not at the centre of the dream though he as yet does not know this. But his claim is more than an assumption, really a reasonable working hypothesis at least until contradicted by better OEG or evidence. The claim is perhaps an important one since (H77) he goes on to claim that dreams help to self-organise behaviour and even more relevantly, to bring about a sense of self, so again we are coming near the hard problem (if any). It is relevant to our A series as well as by now it looks as if the A series may be deeply involved with the self and any sense of self that the brain or mind may have. Different people seem, as we have already seen here, to have their own different senses of past, present and future.

(H85-6) makes it plain that sleep deprivation can literally kill, and not in some 'hysterical' way, but by means such as a breaking down of thermoregulation, and the effect has been clearly exhibited by rats. So there is undoubtedly a mind-body nexus of some sort and (H87) suggests that, roughly speaking, sleep 'tidies up' the mind and at the very least performs some operation in some way analogous to disc scanning and defragmentation on a computer. That would seem to be proved by the rat experiment just referred to.

It is believed (H90) that unpleasant phenomena like nightmares are caused by the limbic brain.. (H92) carefully distinguishes night terrors, usually found in from unpleasant dreams or nightmares. Inhibition dampen the motor system during normal sleep so that we do not sleepwalk

(H101) Going to sleep enables brain activation processes analogous to delirium and Alzheimers.. (H109-133 including Fig 9) shows the part of the brain involved in REM and NREM sleep. We should also in mind later work (Dolan et. al. (2005), Cosmides et al. (2004), Vogeley (2001), Franklin (2005)) but even with the use of more modern methods (e.g. PET instead of fMRI), roughly the same brain areas are still important as in Hobson's Fig 9 as far as I know. In particular the dorsolateral prefrontal cortex tends to be de-activated during sleep and the anterior cingulate and some other regions activated in REM sleep.(H113) fills in some details and tries to show (reasonably well) why dreams are what they are. So we tend to conclude that the multimodal sensory cortex and the deep frontal white matter are ESSENTIAL for dreams.

Obviously, perhaps, what is important is that white matter connections within the brain are what count and the above areas act as staging posts or relays. Now it is well known (H119) that direct temporal lobe stimulation can obtain dreamlike states and furthermore Stickgold's work (H120) relates dreaming to memory. It is also well known that Stickgold (2005) has made it clear by his famous Tetris experiment referred to in his review mentioned above, that dreams can be manipulated beforehand. Thus dreaming relates clearly to memory. Also experiments can clearly allow us to alter dreams before they occur, and we roughly know why. Now if our personal observations extend in some way into the future as well as into the past by means of the A series, it may be also possible to create dreams for us from the future. I suspect that this is too gross a way to look at the problem or situation but it needs to be carefully tried by clearly devised experiments. No reason why people cannot be persuaded to dream of Tetris prior to being instructed in it, I suppose. But I am here reminded of Isaac Asimov's spoof article on 'thiotimoline', a fictitious compound which was said in the article to have carbon bonds extending into the future. As Asimov said, it was not doing to be possible to 'fool' the substance by later setting up the experiment in the future in such a way that the substance could never move to a position where it had already been pulled by its bonding (the bonds being extended to the future). On such matters I refer to my earlier article of 7th October 2005, where I quote Norman Swartz (1993). Also the literature contains a large number of philosophical articles on the so-called 'grandfather' paradox which in essence is very similar. We need a carefully designed experiment which will not confound or conflict with the subjects pre- dream conceptions and then carry out some exercise like teaching the subject Tetris AFTER the experiment. Tetris itself for this use may have problems (e.g. most people know it already) and in my experience, people who do these kinds of experiment go in there with a firm idea that the experiment will (or possibly will not) give useful results. It is going to be necessary to be both utterly fair and thoughtfully calculating in the extreme in designing and carrying out the experiment, not an easy task. Obviously if a relatively simple experiment like the above does not work, a more complicated strategy my be employed bearing in mind our physical knowledge of the subject. In particular it is not immediately clear that a simple visual notion like dropping items would transfer in exactly this manner, as really a more conceptual though non-ambiguous concept might work better.

(H121-2) Pursuant to the above, REM sleep (and possibly also) seems to help with or to consolidate learning. Further just as there seems to be a 'forward sleep window' relating the amount of REM (and possibly NREM) sleep to learning, so too there may be a 'backward sleep window' relating REM (and possibly NREM) sleep before learning to the subsequent learning. The A series certainly extends to the past, as far as we can see, but whether a direct and easy extension to the future can be found could be another matter, but the comments in Section 1 seem to make it clear that such an extension is there, in some way.

Dreams also seem (H123) to have a strongly emotional character, and whilst we may not want to go as far as Revonsuo with regard to taking anxiety as being a prime function of dreams, clearly many dreams that are particularly noticed and frequently appear to be repetitive are anxiety dreams. This interesting feature could also be made part of an A series experiment. (H123) also mentions that, during a dream there seems to be a 'timeless' quality, i.e. we normally do not remember during a dream, for example, that someone in the dream has just died. This is a curious and interesting feature of dreams (if it turns out to be more or less universal) and certainly ties in with the present approach. Indeed times, places and people all (Hobson says) tend to be plastic. And subjects (we might almost say 'process subjects' as in exam dreams for example) tend to be more important than clear previous fact. Sleep clearly helps to concretize certain tasks (H124-5) as has been showing by the subject carrying out the task before and after sleep. Subsequent to the sleep, the task is frequently carried out better and/or more successfully. It could be that this effect also can be shown to work in a retro experiment as discussed in the previous paragraph. (H125) These effects could be highly replicable if carefully set up and even proved further by brain imaging experiments!

In this regard there is the added bonus for the idea of A series involvement that disrupting REM or patterns prior to a task tends to make a subsequent relevant task reach less efficiently performed. We know that already, without the need for an experiment simply to prove the fact. So the idea of A series involvement is beginning to look a natural part of phenomena, rather then a gratuitous half-baked addition. There is thus some ground for optimism.

(H129-131) suggests important tests for a more precise model of dreams of the future. i.e. if we can correlate somehow autobiographically to past processes, as Hobson has attempted to do, perhaps we can also correlate to future learning process (such as even learning Tetris, though as explained that is probably not ideal).

(H136) is still trying to come up with answers which will resolve the binding problem. Dreams may well provide many clues not only to the binding problem, but to the hard problem. (H139) touches on qualia and dreams and how self-activations of the brain relate to qualia. At this point he finally starts talking about lucid dreaming (H140-3) which I suppose can be taken as a cross-over position between sleep and dreams - which of course should elucidate a lot of things including the hard problem and the binding problem. In fact on (H147) Hobson gets to the point of saying "The ultimate reality of consciousness (probably) includes, and is strongly based upon, our brain's capacity to create a virtual reality, so close in all its formal details to aspects of waking consciousness as to fool us, almost every time". I am not clear of the full meaning or implications of this strong statement but it may well be one reasonable intimation or interpretation of the facts of the matter.

Section III

I am not in this section suggesting that anything other than Lucid Dream (LD) detection can be carried out (or considered) along with the measurements described herein. I obviously have other reasons, e.g. along the lines, though certainly not identical lines, to the work that both Jeffrey Gray and myself have both done on synaesthesia and functionalism, but I will return elsewhere to such matters. And there is still more news on my synaesthesia experiments, as by now I have done quite a lot of experiments on claimed synaesthetes, and some on hypnosis also.

I also make the point that Hobson, Domhoff and particularly Metzinger (apparently a lucid dreamer) all inevitably have their own views which do not completely coincide and overlap. This section of the work is basically the section based on Metzinger's work. We bear in mind Hobson (2005a) "lucid dreaming is a valid and potentially useful state of consciousness".

There are nowadays many differing kinds of apparatus which can detect whether dreaming is occurring. these help in the induction of lucid dreams and may at some point be usable to tell when a lucid dream occurs. Some principal methods at the moment are changes in the peripheral arterial tone signal (e.g. Watchpat100), IR detection (LaBerge's NovaDreamer and others), and EOG/EEG measurements.

Time Travel in Theory and Practice

The classic approach is possibly EOG with or without EEG. We are currently going to use OpenEEG project. Amplifier boards made up by Olimex in Bulgaria are under $200 for a two- electrode set and the open source software allows a lot of flexibility. Often in the past EEG and OEG measurements have been done simultaneously but by the look of graphs like those in the Hobson (2005) paper we will get away for now with just EOG. However with the OpenEEG system we can expand at small extra cost if we have to. Also, we can even study lucid dreams without any equipment at all. To get started we are working without dream detection equipment until the apparatus is built. Most work has so far been done this way.

Overall Metzinger (2004) seems to have written a long, fascinating book. He has a great knowledge of lucid dreaming and many of his ideas and concepts may well be incorporated in its further studies, without acceptance of all his ideas. For this reason, I have referred to the individual pages in 'Being No One' in brackets as (M...) in the following. I also refer to his paper (Metzinger 2003) "Phenomenal Transparency and Cognitive Self-reference" as (MA...)

As he points out (M57) "Simulations are always embedded in a global representational context". Leaving aside LD. he points out (M172) that there seems to be a graduated variation between LD and non LD. ie from completely phenomenally opaque to not so opaque for the global representational content. On (M256) it seems that the flashing of LEDs during or approaching REM sleep is likely to influence the dream (which for example could then be about flashing police car lights) as well as perhaps to introduce lucidity.

Perhaps somewhat naively, I refer here to Block's inverted earth model and the phenomenon of tracking also (http://plato.stanford.edu/entries/qualia/) ["The suggestion that tracking is teleological in character, at least for the case of basic experiences, goes naturally with the plausible view that states like feeling pain or having a visual sensation of red are phylogenetically fixed (Dretske 1995). However, it encounters serious difficulties with respect to the Swampman case mentioned above.

On a cladistic conception of species, Swampman is not human. Indeed, lacking any evolutionary history, he belongs to no species at all. His inner states play no teleological role. Nature did not design any of them to do anything. So, if phenomenal character is a certain sort of teleo- representational content, then Swampman has no experiences and no qualia. This, for many philosophers, is very difficult to believe."]. [also "The general topic of the origins of qualia is not one on which philosophers have said a great deal."] We may here be led to issues of cladistic parsimony also.

So at this point we are really speaking, from the point of view of some explanation of what's going on, of the relevance of cladistic parsimony to an interpretation of what goes on in terms of the mathematical physics which we may represent as the alternative worlds of the lucid dreamer. In short we could say that there are a number of lucid dream worlds, at this point analogous to the many worlds of quantum theory but not in our case NECESSARILY an infinity (denumerable or otherwise) of them. We can look on this very simply as rather analogous to the 'many worlds' of a Boltzmann gas and the crux may become - what are our results this way and what do we thus achieve. We seek merely results, not a TOE nor even a 'best at this time' model. But we would like an intelligible explanation and a way to proceed. We seem to be coming up with the use of further parameters or worlds to get an interpretation or results for lucid dreaming. In short a really useful A series explanation of the world, incorporating such factors could be more difficult to derive than it looked at first, but it is interesting enough and not out of sight. And we are certainly not stuck with representationalism as a sole cure or set of guidelines to our situation.

Further, on the matter of the principle of parsimony: It cannot be denied that dreams exist and most believe that lucid dreams also exist. This means that we are almost mathematically obligated to create a useful set of parameters to describe them, a set of state spaces or some other construction. So there is no question of reasonable Occam's razor suggestions at this point. These state spaces or whatever they are can also be contacted by any meaningful physical world so they cannot be ignored either. Given the existence of an A series, these states extend into the future as well as the past but anyway this would be the case for the B series, which is likely to be far less liberal and flexible in potential interpretations IMO however, constrained as it is by Newton and relativity in its present form. As and when we think we understand more details, more detailed descriptions can follow. This gives us more genuine scope than forcing newly described phenomena into an old outline.

Metzinger tries (M256-7) to work out how outside stimuli (or possibly other sources) could effect the course of dreams, and then tries to explain the likely reasons for dreams, in the normal terms of neuroscience. He claims his conclusions may show a parallel with Anton's syndrome and the Charles Bonnet syndrome, interesting but I remain to be convinced.

However he then suggests (M260) ways to extract further information from

283

dreams, e.g. by training colour perception.

I would have thought that further explanation of at least lucid dreaming might be obtained by operant conditioning, either in dream state or conscious state or both, going further than the efforts of LaBerge.

He then, also on (M260) goes further by suggesting a subset of three nonadaptive states, pathological consciousness, machine consciousness and lucid dreaming. In this way, it seems to me, he is almost placing human consciousness in the same bracket as a so-called 'machine consciousness'. Indeed, (M27) he raises the idea that consciousness could supervene on properties which have to be individuated in a more teleofunctionalist sort of way than simply occurs on classic biological properties. To extend the analogy, (M41), he compares a thirsty blindsight patient to a null Turing machine. An interesting and fair comparison.

Further (M199) he says truly conscious entities should be expected to have embodied goal representations which are emotionally grounded. It seems to me that he is looking for 'tracking'.

(M325) actually says we are not things, but processes, and at this point the apparently essential concept of time (probably A series time at that) arises. For time we need consciousness. I'm afraid that here he pre-empts some comments which might have been made to the effect that this could require some kind of autobiographical memory by referring (correctly) to cases of Cotard's syndrome! He also mentions the dangers of dynamic self-consciousness.

He states (M362) that what is needed is a computational model of phenomenal self-consciousness that directly correlates with content and with causal role. To me at this time it all looks as if it could be more A series than B series.

Damasio (M382) has taken the view that consciousness and emotion are inseparable, but wakefulness can be separated from consciousness. Metzinger says (M383) that conscious feelings are historic entities. In essence he is saying that they are not fungible, as I see it.

Metzinger claims (M580) that you can only refer yourself linguistically to your own self-model. Maybe that is good enough for a machine representation anyway leading to a mathematical or physical discussion or model, leaving aside any problem in actually obtaining machine consciousness.

Time Travel in Theory and Practice

A state space (M619) topologically equivalent to the consciousness of human beings could be useful to determine what consciousness is like. Metzinger suggests that such a thing could be obtained by hardware directly. It could also be approached mathematically through software and psychological experiments of simulation IMO. In fact it could even be vaguely like what Vaughan Pratt (2005) has already tried to do for Chu Spaces in ratmech.pdf.

Metzinger points out that dreams may not be entirely the result of some 'internal randomisation' process, whether or not they have a good reason to occur. He looks at dreams as high-level Rorschach tests (M261). The 'internal fairy tale'.

He sees dreams as not epistemically blind, empty artifacts but as an exclusively internal reality- model.

Necker cubes (MA360) maybe illustrate entry into "parallel universes"- or could be set up as a model of an array of objective states, somehow coalescing into one rather like the Boltzmann statistics. To say the effect is quantum, or indeed even to define it as I imply above, could if we are not careful pre-empt the nonexistence of the A series and hence lead to problems. I DO THINK THE PROCESS MUST BE CORRECTLY CATEGORY-THEORETIC at this point.

Metzinger refers to the effect as 'sensory opacity' (MA353). He also compares it with the sensory opacity in lucid dreaming. So it seems to me that, whether neurological or not, we cannot necessarily liken it to neurological effects like say the Capgrass syndrome [at least immediately, clearly and in the same way] which do not evoke the same characteristics AFAIK. We could read (MA360) for further clues. As Metzinger seems to say, the effect is at a different [nearer 'in brain' than 'outside brain'] level than drug-induced hallucinations, and to my mind, is simply a characteristic, like lucid dreaming, of normal states of consciousness. As I have tried to point out above, these effects may well exhibit the A series behaviour more accurately than the B series behaviour of the brain.

(MA360) With say LSD rather than LD (lucid dreaming) "Typically, the subject will immediately have doubts about the veridicality of her experiential state, cognitively "bracket" it and take back the "existence assumption," something which effortlessly goes along with visual experience in standard situations. (Metzinger's) claim is that what this subject becomes aware of are earlier processing stages in her visual system. Visual pseudo-hallucinations –

the breathing patterns on the wall – are such earlier processing stages." Thus Metzinger claims LD are (probably) more internal than LSD say. But he does add "Of course, complex hallucinations, which are fully transparent and in which the experiential subject gets lost in an alternative model of reality, do exist as well (see, e.g., Siegel and West 1975)."

"More importantly, however, the paradigmatic examples for fully opaque state-classes are deliberately initiated processes of conscious thought. In these cases we really experience ourselves as deliberately constructing and operating with abstract representations, ones that we have generated ourselves and which can, at any time, turn out to be false. We are cognitive and epistemic agents at the same time, as thinking subjects actively trying to achieve an expansion of knowledge. In particular, we are introspectively aware of earlier processing stages, namely the formation of thoughts. COGNITIVE REFERENCE IS PHENOMENALLY OPAQUE.(my caps)". TO MY MIND THIS SUGGESTS THE IMPORTANCE OF LD. We must bear in mind that Metzinger apparently is a lucid dreamer and probably therefore in some ways an enthusiast of the idea as he experiences LD himself, FWIW. We could argue that it is therefore natural that he describes LD in the following words: "The best and most basic example for an almost fully opaque, global phenomenal state is the lucid dream (see LaBerge and Gackenbach 2000; Metzinger 1999, 2003).".

Ironically, bearing in mind his current essay "Being No one" we find this raises the question as to whether Metzinger is a drug abuser, as R.P. Feynman seems to have been at about the time that he seemed to have claimed, perhaps abnormally, synaesthesic ability. However my own personal experience, as neither a normally lucid dreamer (nor a drug abuser at any time), is that his LD story seems to hang together anyway. I have few dreams, lucid or otherwise but it sounds OK as a fair hypothesis. And it seems a little more easy to satisfy Popper's principle in the LD case than it does for say the Perky effect and many other psychological experiments when you get to long term plausibility.

Phenomenal transparency and cognitive self-reference. We can therefore consider the many worlds of lucidity as in some ways existing per se and hence deal with them in some sense mathematically, but in some ways they may exhibit the A series more clearly than the B series, or at least one version of it.

How do we proceed ?

Well, Hobson (2005a) takes a slightly different line but we can tie in with it:

He says

"As a first step we would need to quantify the phenomenal characteristics of, let us say, three distinctively different states of consciousness: waking, dreaming and lucid dreaming. How can the phenomenology of these three states be reduced so that it is tractable? Certainly not by focusing on such valid but unworkable aspects of consciousness as transparency. All three of the states of interest have this philosophically celebrated quality! That relegates transparency and many other Metzinger constraints to empirical uselessness.

I suggest that if we take a formal approach to the cognitive quality of the three states we can begin to get somewhere. All of the three conscious states of interest are brain activated states but the EEG is too insensitive to distinguish between the specific activation patterns. Brain imaging can do so however.

Hence it is clear that, compared to waking, dreaming is characterized by activation of most brain regions to the level of waking. In REM sleep some brain regions are informatively more active than in waking. They include: the pontine brain stem which is hypothetically responsible for the endogenous brain activation and the pseudo-sensory stimulation that results in the visuomotor hallucinosis of dreams; the limbic system, particularly the amygdala and the temporal and deep frontal cortical regions to which it projects (which is hypothetically responsible for the hyperemotional and remote memory enhancement of dreams; one cortical region and the parietal operculum, which is involved in visuospatial integration and which may therefore help us understand the remarkably faithful simulation of the outside world in dream consciousness.

But another cortical region, the dorsolateral prefrontal region, is conspicuously less activated than in waking. This specific deactivation may constitute the physical substrate of the cognitive incapacity of non-lucid dreaming. The dorsolateral prefrontal cortex is thought by many cognitive neuroscientists to constitute the physical basis of such executive functions as; working memory; directed thought; self reflective awareness; and critical judgement. Since all of these executive functions are weakened in non-lucid dreaming it is reasonable to propose that it is the underactivation of the dorsolateral prefrontal cortex that causes us to have poor recent memory within and after dreaming and to believe uncritically that we are awake when we are in fact asleep; and to fail to think logically or direct our thoughts when we are dreaming. These journal features can be defined and measured as first person data."

Jackendorff (by Gray, 2002): "We should also not lose sight of Jackendorff's important 1987 book. This gives a searching analysis of the level at which informational structure enters consciousness. Jackendorff proposes an 'intermediate-level' theory of consciousness. Roughly, this holds that the contents of consciousness reflect informational structures derived from a combination (within each perceptual modality) of bottom-up and top-down processing. Jackendorff argues that one is not normally (and perhaps never) aware of sensation unaffected by conceptual interpretation (cf my comment above on the snare and delusion of sense-data), nor of pure conceptual structure, but only of an admixture of the two that optimises the fit between them. This formulation is surely in general true. It is impossible to bring to conscious awareness (or so I find; see the discussion of the square root of -1, above) the pure conceptual structure represented by, say, the string $7 + 5 + 12$ without this taking the form of either seeing or hearing the string, whether by means of externalised stimuli or 'in my head'. Mangan, in contrast, claims that there are exceptions to this rule. In particular, he claims, the pure conceptual structures that can be expressed in English as 'this is familiar', 'this is unfamiliar', 'this is right' or 'this is wrong' can be consciously experienced in the absence of qualia: as 'naked concepts', one might say."

We could also look at Fodor's work - But this 'module' approach is maybe not exactly what we seek.

At all times the Wason restricted set work of Margolis and others should be appreciated too, whether we take a view somewhat similar to Carruthers or not. (That:"The mind is a system of modules shaped by natural selection").

Well as far as we can, we shall for the moment stay with Metzinger, LaBerge and Hobson and also proceed to use operant conditioning on lucid dreams and to relate our work to the A and B series and to category theory.

Sketching out further results which might be obtained.

I will give one fairly extreme example of the possible scope of the present method, and where it actually already works. I show it working below.

This concerns precognition.

Now as far as I know precognition is an idea which is mainly disregarded by established science and considered rather outre, along with PK and ESP. I want to say, really want to say, that this disregard sounds reasonable. I bear in mind all the published work I know of in both of those two fields.

Now that does not stop me thinking about precognition.

It seems to me that precognitive results obtained by dreams seem to fall in much the same bracket as PK and ESP, at first glance. For example there is no statistical proof or other scientific proof whatsoever of any real cases of precognition.. And to my knowledge, practical proof to the extent that any lucid dreamer ever became extremely wealthy from stock market predictions is also unknown (I am not thinking there of relatively short term successes, but the equivalent of persistent and continually verifiable wins at roulette).

However we need to bear in mind that lucid dream precognition would be likely to have certain snags or problems associated with it. Generally speaking we cannot assume that the overall mental universe of the dreamer before the dream and after the dream will be the same. The dreamer may have left one universe and gone to another one, either during his dream or after it or both. If the dreamer did that, we ask ourselves whether he will return to the same physical universe after the dream or have been in it whilst experiencing the dream. To say all that is not really too bizarre, particularly as some may wish to interpret the strange worlds as simply a mathematical phase space. And certainly a 'cancellation of infinities' approach, if indeed eventually adopted to remove so called 'phase spaces' is almost a hallmark of modern theoretical physics.

Now if the dreamer is, for example, not in the same universe during the dream he will not have experienced results which will necessarily give a good or useful predictive pattern. Suppose for example he dreamt event $A(d,U(v),0)$ at time $t=0$. Now clearly if the dreamer is in Universe $U(w)$ the equivalent event which he could be expected to have precognition of for time f is $A(w, U(w), t(f))$ This is unlikely to be necessarily $A(d,U(v),0)$. And we do not know that the theory will be that simple, for example the event dreamt may be a random example of the events $A(d,U(x),0)$ where x has many values or could even be a multivalued function or averaged function of a number of values.

So how can we ever think rationally of precognitive dreams in such cases? Well, to begin with there are as above, fairly clear reasons that the above scheme will give exactly what we are finding on the ground. That is to say, no precognitive results on an obvious statistical basis that we can determine but a number of dreams rather like what Jung might have called 'great dreams' which sometimes sound like striking examples of precognition but don't cut with statisticians. [According to Jung, our subconscious reveals its message as a 'Great Dream'. Usually starting with simple images of the dreamer's personal life, the subconscious proceeds to completely take over and project selected symbols of the collective unconscious http://www.lifepositive.com/Mind/paranormal/dreams/dreaming.asp]

So one example of a predictive dream might well sometimes be the 'great dream' but of course we will not have statistical evidence at the present stage. So dream research is necessary to probably sort out (1) which if any dreams are predictive (2) have they any particular qualities in common. I suspect that something less simple than the usual sorting through large numbers of accumulated statistics looking for a breakthrough would be necessary and indeed we have already got a small payoff as explained above. Any such study has certainly the positive aspect that we have indicated why it may be quite correct that positive clear cut results cannot be easily expected, and that is in line with our thinking here.

We certainly do not claim that the above is the only way forward, in fact extending the results of Section II looks far easier.

Appendix

A few further notes about procedure from Metzinger's book. Much other work, by authors including Revonsuo, LaBerge etc. is available.

(M304) He claims that the content of the Phenomenal Self Model (PSM) will change during the transfer from a normal to a lucid dream.

(M497) On OBEs. These are frequently accompanied by the subject seeing their own body as if disconnected from them, without multicentring or decentring the state of consciousness. The locus of consciousness is typically the etheric double. Some would say OBEs are a subset of lucid dreams.

(M529). Start of the important section. 7.2.5 of the book. At least this section till the end of the chapter needs reading.

(M534) He indicates that fixating visual objects terminates LDs, reliably.

(M536) Factors conducive to LDs.. (a) High level of physical activity during the day. (b) Heightened affective arousal during the day. (c) Short interruptions of sleep, including short activities of waking consciousness preceding the relevant REM phase.

The PSM seems to be anchored in the brain as there is a direct and reliable relationship between the direction of polygraphically recorded eye movements and the gaze shifts reported in the lucid dreams.

Experienced subjects can thus allegedly indicate ocularly when a LD actually starts.

References

Bierman (2002) On 's "Quantum Consciousness" site:
http://www.quantumconsciousness.org/views/TimeFlies.html
[According to Bierman "Dean Radin and Dick Bierman have performed a number of experiments of emotional response in human subjects. The subjects view a computer screen on which appear (at randomly varying intervals) a series of images, some of which are emotionally neutral, and some of which are highly emotional (violent, sexual....). In Radin and Bierman's early studies, skin conductance of a finger was used to measure physiological response They found that subjects responded strongly to emotional images compared to neutral images, and that the emotional response occurred between a fraction of a second to several seconds BEFORE the image appeared! Recently Professor Bierman (University of Amsterdam) repeated these experiments with subjects in an brain imager and found emotional responses in brain activity up to 4 seconds before the stimuli. Moreover he looked at raw data from other laboratories and found similar emotional responses before stimuli appeared. Professor Bierman presented these findings to the recent Tucson conference." This is Bierman's view as of Tucson (2002). Obviously Domhoff (2000) would not seem to have formed the view that the effect occurs in dreams. Indeed. in Domhoff (2000) there seems to be included a blistering and certainly not wholly unwarranted attack on ESP, precognition, and all that.

Bierman also says: 'There's plenty of evidence that time may run backwards,' says Prof Bierman at the University of Amsterdam. 'And if it's possible for it to happen in physics, then it can happen in our minds, too.' In other words, Prof Bierman believes that we are all capable of looking into the future, if only we could tap into the hidden power of our minds. And there is a tantalising body of evidence to support this theory. ' ???]

Blackmore, Susan (2006), http://www.susanblackmore.co.uk/ , http://skepdic.com/esp.html , and many references by others. On the Wiseman (2004) controversy and similar matters, Domhoff (2000) is probably enough to cover it for now. Bostrom, N. (2006). http://www.nickbostrom.com/

Carruthers, P. , (2005), http://www.philosophy.umd.edu/Faculty/pcarruthers/Shaped-modules.htm

Cosmides L. et al, (2004) 'A Theory of Autobiographical Memory: Necessary components and disorders resulting from their loss' Social Cognition, Vol. 22, No. 5, 2004, pp. 460-490

Deutsch, D. (1997), "The Fabric of Reality", 81-2, Penguin

Dolan R.J., et al. (2005), 'Noradrenergic Modulation of Emotion-Induced Forgetting and Remembering ´ The Journal of Neuroscience, July 6, 2005, 25(27):6343– 6349 , 6343

Domhoff, G.W. (2006), "DreamResearch.net Dream Library" numerous articles, http://psych.ucsc.edu/dreams/Library/index.html

Domhoff, G. W. (2000). Dreams and Parapsychology. Retrieved January 9, 2006 from the World Wide Web: http://www.dreamresearch.net/Library/domhoff_2000c.html

Domhoff, G. W. (2002). Using content analysis to study dreams: applications and implications for the humanities. In K. Bulkeley (Ed.), Dreams: A Reader on the Religious, Cultural, and Psychological Dimensions of Dreaming (pp. 307-319). New York: Palgrave

Dretske, F. (1995) "Naturalizing the Mind", Cambridge MA, MIT Press.

Franklin M.S., Zyphur M.J., (2005), 'The Role of Dreams in the Evolution of the Human Mind' http://human-nature.com/ep/articles/ep035978.html

Gray J., (2002), Psyche, 8(11), October 2002

Hobson, J. A., (2005) Nature, Vol 437,27 October, 1255

Hobson, J.A., (2005a) PSYCHE 11 (5), June 2005

Hobson, J.A. (2005b), "13 Dreams Freud Ever Had", 83, Pi Press.

Hobson, J.A. (2002), "Dreaming: An Introduction to the Science of Sleep", Oxford University Press.

LaBerge S.,Gackenbach J. (2000), 'Lucid Dreaming', in Cardena, Lynne, and Krippner, "Varieties of Anomalous Experience", American Psychological Association.

Metzinger, Thomas, (1999, 2003), many papers, some mentioned in Metzinger (2004).

Metzinger, Thomas, (2004) 'Being No One', MIT Press Paperback.

Metzinger, Thomas, (2003) "Phenomenal Transparency and Cognitive Self-reference", Phenomenology and the Cognitive Sciences 2: 353–393.

Newhousenews (2006) Examples are cited in many places including
[a] http://www.newhousenews.com/archive/seeman040704.html

b) http://faculty-gsb.stanford.edu/millerd/docs/1989jpsp.htm

c) http://www.infm.ulst.ac.uk/~paul/pubs/aaai94.ps
"given any particular picture, for example an abstract evocation of the dawn, two agents would not necessarily produce equivalent propositional descriptions of it. However, it does not follow that the methods they are using to translate between representational mechanisms are incompatible. For example, two people may produce completely conflicting descriptions of a scene, and yet still be willing to accept an argument as to why the given description is valid in the context of the other agent. "

d) http://www.its.caltech.edu/~jbogen/text/mentali.htm "both the left and right hemisphere may be conscious simultaneously in different. even mutually conflicting. mental experiences that run along in parallel"

e) Then there is the classic elementary psychology experiment where a student acts as an assailant in a psychology class, apparently shooting the professor and the observing students give varied and genuinely believed accounts of the incident.

Raison's reason as cited in

"http://www.newhousenews.com/archive/seeman040704.html "cited in a) above, may apply. Raison even gives a personal example ""Many years ago, I was a journalist working for a newspaper," Raison said. "I was back in the composing room, and I heard a ruckus. There was a guy waving a gun around and pointing it at his ex-wife's head."
When the employees met later to discuss the incident, their stories differed widely, even on fundamentals like which door the attacker used to flee.
"It was amazing," Raison said. "It was like we all saw a different thing. I didn't understand how one person could be so wrong.""]

Pratt, V, (2005), http:// boole.stanford.edu/pub/ratmech.pdf

Stickgold R., (2005) Nature Vol 437 , 27 October, 1272

Swartz N., (1993) http://www.sfu.ca/philosophy/swartz/time_travel1.htm

Tucson (2002) http://www.conferencerecording.com/newevents/tsc22.htm

Vogeley, K., Bussfeld, P., Newen, A., Herrmann, S., Happé, F., Falkai, P., Maier, W., Shah, N. J., Fink, G. R., and Zilles, K. (2001). Mind reading: Neural mechanisms of theory of mind and self-perspective. NeuroImage, 14: 170-

Wegner, D. M. (2002). For example., "The illusion of conscious will." Cambridge, MA: MIT Press. Wiseman (2004) http://www.victorzammit.com/archives/Aug2004.html

7.10 Work in Progress on application of dynamic systems theory to the A series (1)

At this point there seem to be a variety of ways to proceed. The FitzHugh-Nagumo (FHN) model, as used by Longtin (2002) and others, could perhaps be very useful - we'd have liked to have gone down that road - but does not seem a sufficiently developed approach at the present time (Note 1) and as it stands could add complications without adding enough merit.

The approach, roughly defined by Lange (2001) and indeed Guastello (2002) also carries problems. Something a bit more basic, beginning with the work of Sprott (2005) and Gottman (2002), is a more basic approach. There seems no reason why it cannot be a useful line to follow, bearing in mind the analogous success of the Kuramoto model and then extended work along the same lines. As Strogatz (2003) points out, that very basic model and indeed almost too simple a model did lead to further breakthroughs. So something basic is what we will try: At this point, in our hearts is phase locking of a leaky integrate-and-fire model - at the very least - but we stick with looking for a simple and fairly descriptive model at least for a start. Like Winfree, perhaps, we are looking for early experimental success.

Dreams were first made predictable by the work of Stickgold (2005) who showed us how to actually create dreams in advance. Whilst we would like to consider them to be dreamt prior to their creation as I seem to have already done in accord with possible expectations (Yates, 2006), the neurology should have similarity so we can look at the Stickgold work first.

I am waiting to obtain my copy of Stella and may use it to explore the present situation in more detail [in practice we used Madonna]. I have already looked briefly at Rinaldi's (1998) Laura-Petrach model (Note 2), obtained graphs, and thus note that suitable comprehensible models are possible.

References

Gottman J.M., et al., (2002), "The Mathematics of Marriage", MIT Press , ISBN : 0-262-57230-3

Guastello S.J., (2001), "Nonlinear Dynamics in Psychology", Discrete Dynamics in Nature and Society, Vol. 6,11-79 ; "Progress in applied nonlinear dynamics: Welcome to NDPLS Volume 8." Nonlinear Dynamics, Psychology, and Life Sciences, 8, 1-16.

Lange R., Schredl M., Houran J., (2001), "What Precognitive Dreams are Made of", Dynamical Psychology ; http://goertzel.org/dynapsyc/2000/Precog%20Dreams.htm#_ftn1

Longtin (2002) Fluctuation and Noise Letters, Vol. 2, No. 3 (2002) L183-L203

Rinaldi S., (1998), "Laura and Petrarch: An intriguing case of cyclical love dynamics", Siam J. Appl. Math., Vol. 58, No. 4, pp. 1205 - 1221

Sprott J.C., (2005) , "Dynamical Models of Happiness", Nonlinear Dynamics, Psychology, and Life Sciences , Vol. 9, No. 1, January; and others.

Stickgold R., (2005), "Sleep-dependent memory consolidation" , Nature, Vol 437, p1272

Strogatz S., (2003) "Sync", p60 and elsewhere, Penguin Books, UK. ISBN: 0-141-00763-X

Thomson Gale publ. (2005), "The Rise and Fall of Catastrophe Theory" from "Science and Its Times", 2005-2006 Thomson Gale

Weckwesser W, (2006)
http://math.colgate.edu/~wweckesser/solver/LauraAndPetrarch.shtml

Yates J., (2006), "Can dreams predict the future ?".
http://ttjohn.blogspot.com/2006/04/do-we-dream-of-future.html

Notes

1. The following quote gives one Encyclopedia style summary (Thomson Gale, 2005) of the recent situation on catastrophe theory: "Despite the initial acceptance of the theory, it eventually became controversial. The number of variables involved in a discontinuous process must be small in order for catastrophe theory to model it with any accuracy. In the real world, however, especially in inexact sciences such as biology and sociology, these conditions rarely occur. One less than practical application of catastrophe theory involved its use to model the escalation of hostilities between nations. The variables used were threat and cost. It was argued that catastrophes—in this case, sudden attacks or surrenders—would occur when threat and cost were both high. Although such a model might be used to describe theoretical nations in very general terms, many more variables come into play when real people and real nations are involved. Therefore, such a model could not be used to make predictions of any practical value. Catastrophe theory was also applied with varying degrees of success and failure to social topics ranging from the stock market to prison riots to eating disorders."

"Varying degrees of success and failure" is probably the key term. Results depend very much on the particular model and the degree and variety of accuracy expected from it. The attitude of mind of a theoretical chemist rather than that of a fundamental mathematician or even psychologist is probably more appropriate in very general terms, but all techniques (including catastrophe theory in that name or some other) available to the problem could need to be applied. In that sense we are still in accord with workers such as Winfree or even May, up to a point.

2. I used the stimulus made by the computer in producing the dream to place Laura as the conscious mind and Petrarch as the dreaming (unconscious) mind of the same person. This could mean that beta1 is not zero (when it is, you do not seem to get a phase diagram for the values considered). I mainly used $L(0)$ as 2 with delta as 0 and lowered the coupling between 'conscious' and 'unconscious' by bringing beta1 down to 0.1 .

The periods used were not 'years' but brief periods like days or weeks. In fact the interesting results for me occurred mainly within the first three 'periods'. I'm not too clear how gamma is best interpreted in this rough model and did best leaving it alone.

Also raising beta1 to 10 (strong coupling between conscious and unconscious) was tried and increasing the rate of fade for Laura to alpha1 as 100 and a quick fade for Petrarch as alpha2 as 10 . Neither gave an obvious loop.

I will not show the graphs as readily available mathematics programs like Madonna can produce graphs. Indeed some graphs can be readily obtained using a web solver and for Laura/Petrarch I would mention Weckwesser (2006). Values such as alpha1,2,3 as (3,1,0.1) and beta1,2,3 as (10,10,10) gamma as 1.0 , delta as 0 , AL as 2.0 and AP as - 1 are intriguing for example. What is needed is a slightly more precise approach which I am presently considering.

7.11 Work in Progress on application of dynamic systems theory to the A series (2)

Choice of software.

For much of this work it was found that the Macey & Oster program "Madonna (8.3.11)" (Note #1) provided a simple way to give models of real situations and I used this sytem rather than "Stella (3.0.7)" which is rather similar, and has an excellent written instruction handbook (in Hannon (1997)) and numerous example. Stella was used occasionally too, but the models are nearly interchangeable in practice here. Another model I gave careful consideration to using was the G.A. Korn program "Desire", but most of what we wanted to do so far seems possible in Madonna. The copy of Desire which I obtained did not come with the Windows distribution. I do in fact have a running Linux box, but as the work had been done so far in Windows it was decided not to use the rather more complex but possibly more flexible Desire at this time. One point of note was that inverse Fourier transforms, for example, if required can be obtained from the quite flexible Algebrus, which is readily available as a Windows program. For much of this work Windows 98SE has been adequate to date.

Madonna, like Stella, has the advantage that the actual models simulated are produced at the time the program is prepared, rather than just leaving us with a lot of algebra to report.

Discussion

Madonna makes it very easy to simulate numerous possible models for a system and it is essential to choose one best for the needs of the time. I have tried a great many models and find that a very simple conscious (waking) / unconscious (dreaming) model can give quite good results.

This model is referred to as **N003b**.

In this model shown in N003b diag.clp (see Note #4 to find graphs and equations) the unconscious mind (Romeo) is R and the conscious mind (Juliet) is J. The equations N003b equns.rtf can be substituted into Madonna .

Further elaborations are briefly noted later in this piece.

Now this model works well for descriptions of precognitive dreams, using the Sprott (2004) values for a hermit Romeo and an eager beaver Juliet. (Note #2)

For Stickgold (postcognitive or Tetris) dreams we need to vary values slightly from the (Sprott, 2004) values of Note #2 which as mentioned below correspond to a 'hermit' Romeo and an 'eager beaver' Juliet. In fact the example graph [N003b graph S a -1.99 c 1.01 d 1.017.bmp] or in brief N003b graph S uses a slightly more enthusiatic Juliet (with c increased from Sprott's value 1 to 1.01 and d increased from 1 to 1.017) and a slightly less introverted Romeo (with a decreased from -2 to -1.99). e is also varied from the standard value of 1 down to a lower value of 0.6 which permits postcognition (or for these values almost simultaneous cognition which could be lowered further if appropriate).

Now this seems to be a fairly good model to be going on with, and the relative positions of red and black lines with respect to each other can fluctuate by parameter change.

Some slightly different models, including one with some random elements in R to simulate random dream effects, are mentioned in Note #3.

Stickgold's results and my extensions to them

Stickgold's dream creation results devolve around the fact that it appears that we have two different memory systems. The hippocampus codes information on events from our lives. The findings suggest that the brain does not go to the hippocampus to get images for dreams, but to the long term, neocortical system. Stickgold uses evidence like the fact that in one series of experiments, three amnesic patients with extensive bilateral medial temporal lobe damage produced similar hypnagogic reports to the control sample despite being unable to recall playing the game, suggesting that such imagery may arise without important contribution from the declarative memory system.

I set up a preliminary model M002 (see Note #4) which contains a representation of a sleeping mind (R), a waking mind (J) and a perturbation (Z) (or 'Tetris or ski-ing game') and the details and parameters need to be filled in and/or added to or altered to give flesh to the model.

Then a large number of other models and parameter sets were tried (Note #3) and mainly we used N003b as stated.

A few neurological details as to obtaining choice of model

A Kahn and Hobson (1993) proposed the use of a simple Verhulst equation model as a starting point. In their equation (2) they formally visualise x as the density of images in a dream report and t as the time. This gives

$$dx/dt = ax - x*x$$

299

or in a more advanced form,

$$d x\_t = (alpha * x\_t - x\_t * x\_t) * dt + F * x\_t * dW\_t$$

where alpha is described by equation (4) of Kahn (1993) as an 'average value', x_t is the stochastic variable and W_t the Wiener process (that is, the basic process of Brownian motion).

B Clearly, remembering and forgetting could be two important features of a model. The second Christos (Christos (1996), Goertzel (1997)) experiment, which uses the Crick-Mitchison (1983) hypothesis (or some similar credible approach) to produce a Hopfield net which (unlike a properly trained Hopfield) produced some positive results.

Sprott (2007) used tanh functions for his neural net model of the logistic map and the same approach could be tried for further work on Christos's problem. Goertzel also has looked at this aspect of the topic of dreams in great detail.

C Flor (1998), for example, uses a simple log-linear model to give the operation of the brain.in terms of task response times.

$$RT(t) = k * t ^ b$$

where RT is task response time, t is the time and k and b are constants. This seems to me to be more or less rather like the use of the common Stephens' formula.

D Hannon (1997) mentions stochastic resonance effects in Chapter 16, and such effects can also be incorporated in a model and in fact I did so in the slightly more complicated SR003A which includes some aspects of the Hannon model in the Romeo or dreaming state. In common with results for many cases where modelling is made slightly more complicated but requires more parameters, so far this does not seem to have really paid off at this level of model making. It might be a way forward at a later date however.

E FitzHugh-Nagumo models. Izhikevich (2007) has a large page of publications, some devoted primarily to the FHN effect and Hasegawa (2006) had a lot of articles both in citebase and in the condensed-matter archive arXiv relating to the FHN and the neocortex and indeed small-world results. I found

Time Travel in Theory and Practice

Sailer's (2007) dissertation contained a useful summary of some of the recent work. Hannon (1997) Chapter 11 outlines how it can all be done but the current situation is pointed out in models like that in section D above (SR003A) I have done just a little introductory work here. Use of more known physical parameters over a range might help, perhaps somewhat along the lines of the way Kahn (1993) approached the matter.

F It may be a good idea to differentiate between the forms of our R and J, so it could mean altering the model so that we may for instance want a Hobson and/or Christos style of model for R and say a Flor model for J. And then to get the parameters using bang-bang or Pontryagin method. This complexity has so far not been needed and I tend, by analogy, to look at the relatively high success of the very simple early Kuramoto models and at the fact that later models seem to add more problems for the number of concrete results obtained., admittedly not inconsiderable in some cases.

Results from Model N003b

N003b is the model considered most appropriate so far.

N003b equns.rtf are the equations for a very simple model.

The diagram N003b diag.clp shows how it works and it is displayed on the graph. Anyone with a Java enabled copy of Madonna can reproduce this system and easily vary parameters.

The small green line on N003b graph.bmp describes a simple impulse or blip, to represent for example a Tetris game, and it is on the same sale as R. The black and red lines are respectively Romeo (call it unconscious mind, dreaming mind or what you will - no complex neurology system or pseudoscientific patter is necessarily implied by the term) and Juliet (conscious) - R and J are both in the same brain.

The a and b values of -2 chosen for Romeo (unconscious mind) are the 'hermit' values (Sprott, 2004). The Juliet values of unity for c and d are the 'eager beaver' values (Sprott, 2004). This suggests that the conscious mind is eager to look and to interact and the unconscious mind appears more like a hermit, (which may well have hidden depths of course). The two of course do interact and in the present model the blip is only supposed to interact with the unconscious mind, or in a dream if we like. These are preliminary choices only to keep in line with earlier work. In fact it was not found that substantial variations from these reasonable choices do much more than add parameters unless there are clear reasons to do so.

N, M and P of course just confirm the there is a real pulse or blip at around time 31, affecting an otherwise blank or 'normalised' mind in the way shown on the graph.

As e is raised from a value of unity through 2 (as shown on the graph N003b e 2.bmp - similar notation for other graphs) to say 10, the time values of R and J peaks get lower and lower, until there are two series peaks for each on the graph, one being dream precognition peaks quite early and the other possible real peaks when the dream comes true. By the time e is 1000 and there is strong interaction, the main peaks occur 'at or around' the time of the pulse again.

Note that if we try e less than unity, at say 0.5 both R and J peaks rise to times later than the pulse. So clearly on this model the intensity of the pulse has to be finely adjusted to allow a 'precognition' effect.

The black line (the dream and/or unconscious mental impingence) rises just before the red line (the observed result) in time - which can be through direct observation or memory of the dream. By altering the parameters the extent of the black line's peak preceding the red line's peak or vice versa can be altered. In fact in Fig 1 they are about equal, as in the case of a dream and its first physical recollection being almost simultaneous - clearly the various parameters allow fine tuning as to their relative positions.

At a zero value of e, there are still peaks at very high time values and these are probably simply explained by model interpretation, model crudity and butterfly effect. At e = 0 there is a very strong butterfly effect as can be readily seen by altering INIT R and INIT J by say 0.001 on this simple and illustrative model. This variation is very clear in an R v J (or phase diagram), and is extremely important here just at e = 0.To put it differently, in the A series if the past, present and future are laid out on the same time line like this, the past present and future may not map adequately onto a B series diagram, so at large times, or at otherwise anomalous times, we cannot expect such a model to necessarily be suitable. We might well feel therefore that in our pseudo A series model, in practice high large or anomalous peaks can fairly be discarded though there is a mathematical reason for them to be there.

On the other hand for very large values of e, or very large interactions, the time values of R and J may not show very great anomalies at large times. In fact they may become rather pedestrian, as they seem to do here at e = 1000.

As pointed out above, if we try e less than unity, at say 0.5 both R and J peaks rise to times later than the pulse, and if e is very much larger than unity the results again seem pedestrian. So clearly on this model the intensity of the pulse interaction has to be finely adjusted to allow a 'precognition' effect. But on this model there is a reasonably wide range of interaction adjustment allowing us to obtain precognitive effects, not necessarily a single point "sweet spot" or cusp.

Conclusions

Up to a point the Dream postcognition model (Stickgold) (Figure 8) and the Dream precognition model (Yates) (Figure 1) speak for themselves. And we can also see that the interesting precognition effect can occur over a serious but possibly brief range of parameter values, as remarked above, and does not necessarily just refer to a singularity or a cusp. Also, the whole idea fits in well with the A series idea as essentially we are only considering a fragment or portion of the time of one individual, and his or her past/present/future.

There is clearly very much more to be said about the use and utility of such a model, which could lead to further awe-inspiring results and further experiments, possibly of a very targeted nature in terms of psychological environment. That is: We are not thinking of yet another general dream survey of the kind which we already have in large numbers, and which indeed are often useful and of great worth.

Much more mathematical explication may be advisable as well as more experiments. Both can occur together. There is a great deal of ongoing experimentation (for example in Garcia-Ojalvo (2004)) from many fronts and I am deeply impressed by the relatively easy blog of Harris (2007, and also 2007a) and likewise by the fine philosophical and psychological comments of Draaisma (2000) also quoted with some approval in Harris's blog.

I will just conclude for the moment in pointing out the relatively easy use of Fourier transforms using the present type of model, perhaps directly or by using the Algebrus program (Note #1) and the possible easy relation of holographic models to the present model in this way. I give some brief but perhaps relevant and important details in Note #5 where I further indicate why the homunculus paradox in our model need not, even carefully bearing in mind

Time Travel in Theory and Practice

Draaisma (2000) on that matter, be of any difficulty with the present approach.

Important Note

The appropriate graphs are essential and thus were included, and anyone who cannot read them on this book can get them from me by email or by post. The coloured versions are on our website ifsgoa.com

References

There are other references located in earlier blog entries but I have tried to include some of the more important ones used here.

Bohm D., Hiley B.J., (1993) "The Undivided Universe" p353,   Routledge. ISBN 0-415-12185-X or on a tribute website http://www.vision.net.au/~apaterson/science/david_bohm.htm#HOLOMOVE MENT

Christos, G. (1996) Investigation of the Crick-Mitchison Reverse-Learning Dream Sleep Hypothesis in a Dynamic Setting. Neural Networks, 9, 427 - 434.

Crick F, Mitchison G., (1983), "The function of dream sleep", Nature 304, 111-114

Draaismsa, D. (2000), "Metaphors of the Mind", particularly chapter 7, Cambridge, ISBN 0 521 65024 0

Flor R., Dooley K., (1998), Noetic Journal, 1(2): 168-173

Goertzel B., (1997) "From Complexity to Creativity", Chapter 10: 'Dream Dynamics' , Plenum

Garcia-Ojalvo J, Elowitz M.B., Strogatz S.H. (2004), "Modeling a synthetic multicellular clock: repressilators coupled by quorum sensing", Proc Natl Acad Sci U S A., Jul 27;101(30):10955-60. Epub 2004 Jul 15.

Hasegawa H., (2006) for example http://arxiv.org/abs/cond-mat/0506301

Hannon B., & Matthias R., (1997) "Modelling Dynamic Biological Systems" , Springer.

Time Travel in Theory and Practice

Harris (2007), http://nine-radical.blogspot.com/2006/11/preview-of-blog-in-early-1990s-our.html ; especially "5. Gems in a junkyard" on holography.

Harris (2007a), a wide range of often relevant papers can be found at http://www.cnel.ufl.edu/hybrid/publication_paper.htm, particularly for example his reference 9

Harris, Nicolelis et al "Ascertaining the importance of neurons to develop better brain-machine interfaces"

Izhikevich E.M. , (2007)
http://vesicle.nsi.edu/users/izhikevich/publications/index.htm

Kahn D., Hobson J.A., (1993) "Self Organization Theory of Dreaming", Dreaming, Vol. 3, No. 3

Prideaux J., (2000) "Comparison between Karl Pribram's "Holographic Brain Theory" and more conventional models of neuronal computation", http://www.acsa2000.net/bcngroup/jponkp/

Pribam K., (2007)
http://www.scholarpedia.org/article/Holonomic_Brain_Theory

Sailer X., (2006), "Controlling Excitable Media With Noise" (Dissertation), edoc.hu- berlin.de/dissertationen/sailer-franz-xaver-2006-03-31/PDF/sailer.pdf , Humboldt-Universität zu Berlin, Mathematisch-Naturwissenschaftliche Fakultät I

Sprott J.C., (2004) "Dynamical Models of Love", Nonlinear Dynamics, Psychology, and Life Sciences, Vol. 8, No. 3, July.

Sprott J.C., (2007), "Neural Net Model of Logistic Map", technical online notes, http://sprott.physics.wisc.edu/chaos/nnmap.htm

Stickgold R., (2005), "Sleep-dependent memory consolidation" , Nature, Vol 437, p1272 ; popularly in Leutwyler K, (2000) "Tetris Dreams", Scientific American, October 16.

Yates J., (2006), "Can dreams predict the future ?", http://ttjohn.blogspot.com/2006/04/do-we-dream-of-future.html ; and many other entries in this blog.

Notes

#1. Madonna : http://www.berkeleymadonna.com/
Stella: http://www.iseesystems.com/ ; a very useful guide book for Stella and also helpful for is Hannon & Matthias (1997)
Desire: Korn, G.A.: "Advanced Dynamic-system Simulation: Model Replication and Monte Carlo Simulation", Wiley, New York, 2007.
Algebrus: http://www.astrise.com/software/algebrus/

#2. Specifically this is the model described in equation (4) of Sprott (2004).page 310. Calculations with the slightly simpler model (1) seem ultimately to lead to similar overall conclusions.

#3. **SR003A**
This includes a random factor during the dream. It does not improve our results technically so far. There are more parameters, not just e but f,k1,k2,k3,F2.

This one has both R and J involved and even e = 100 does not seem to move the pulse much from a sort of 31 on R and J - this is with f = 5

It did not seem to make much difference anywhere f was between +50 and -50and eventually tried e over +100 to -100

So both R and J are involved - but there is not much that is not pedestrian.

Other working notes for a few models are mentioned in this note (Note #3) and it all could be expanded on in a later essay - if there seems need to.

A001btempnew This actually seem to give a meaningful result at only about f = 81 (like e = 81 effectively) and no precog. effect. It also included more terms like a sine wave and a random term in R an a cusp effect. But the pulse was only on J which seems to be one source of less interest for the model, whilst the random (or dream) effect was on R.

N003a which had not connection whatever to the pulse and n e term, also had butterfly like behaviour for small INIT J and INIT R but a good behaviour for slightly larger INIT values, say over 0.001 each i.e. The existence of the pulse does not seem to determine other characteristics than that conncerned with the pulse itself - and we should keep away from very high and very low times on the model, probably.

There were many other models tried including **A001, SR003, SR002, M004**.

#4. Captions to diagrams, and equation details:

Figure 8 DREAM POSTCOGNITION MODEL (Stickgold)
Figure 1 DREAM PRECOGNITION MODEL (Yates)

Figure 1 N003b graph e1
Figure 2 N003b graph e2
Figure 3 N003b e10 graph
Figure 4 N003b e1000 graph
Figure 5 N003b e0.5 graph
Figure 6 N003b e0.1 graph
Figure 7 N003b e 0 graph
Figure 8 N003b graph S a -1.99 c 1.01 d 1.017
Figure 9 SR003A graph Figure 10 N003b diag Figure 11 SR003A diag

N003b equns.rtf is the following:

These are the Madonna equations for a very simple model. The diagram shows how it works and it is displayed on the graph. Anyone with a Java enabled copy of Madonna can reproduce this system and easily vary parameters.

The small green line describes a simple impulse or blip, to represent for example a Tetris game, and it is on the same sale as R. The black and red lines are respectively Romeo (call it unconscious mind, dreaming mind or what you will - no complex neurology system or pseudoscientific patter is necessarily implied by the term) and Juliet (conscious) - R and J are both in the same brain. (The book version is monochrome – colours appear on website.)

The a and b values of -2 chosen for Romeo (unconscious mind) are the 'hermit' values (Sprott, 2004). The Juliet values of unity for c and d are the 'eager beaver' values (Sprott, 2004). This suggests that the conscious mind is eager to look and to interact and the unconscious mind appears more like a hermit, (which may well have hidden depths of course). The two of course do interact and in the present model the blip is only supposed to interact with the unconscious mind, or in a dream if we like. These are preliminary choices only to keep in line with earlier work. In fact it was not found that substantial variations from these reasonable choices do much more than add parameters unless there are clear reasons to do so.

N, M and P of course just confirm the there is a real pulse or blip at around time 31, affecting an otherwise blank or 'normalised' mind in the way shown on the graph.

As e is raised from a value of unity through 2 to say 10, the time values of R and J peaks get lower and lower, until there are two series peaks for each on the graph, one being dream precognition peaks quite early and the other possible real peaks when the dream comes true. By the time e is 1000 and there is strong interaction, the main peaks occur 'at or around' the time of the pulse again. Note that if we try e less than unity, at say 0.5 both R and J peaks rise to times later than the pulse.

At a zero value of e, there are still peaks at very high time values and these are probably simply explained by model interpretation, model crudity and butterfly effect. By e = 0 there is a very strong butterfly effect as can be seen by altering INIT R and INIT J by say 0.001 on this simple and illustrative model. To put it differently, in the A series if the past, present and future are laid out on the same time line like this, the past present and future may not map adequately onto a B series diagram, so at very large times, or at otherwise anomalous times, we cannot expect such a model to necessarily be suitable.

On the other hand for very large values of e, or very large interactions, the time values of R and J may not show very great anomalies. In fact they may become rather pedestrian, s they seem to do here.

{Top model}

{Reservoirs}
d/dt (R) = + x1
INIT R = 0 d/dt (J) = + x2
INIT J = 0
d/dt (Z) = + x3
INIT Z = 0

{Flows}
x1 = a * R + b * J * (1 - ABS (J)) + e * Z
x2 = c * R * (1 - ABS(R))+ d * J
x3 = h * SQUAREPULSE (N,M) - h * SQUAREPULSE(N+ P,M) + h1 *
SQUAREPULSE(N +
h2,M) -h1 * SQUAREPULSE(N+P+h2,M)

{Functions}
a = -2 b = -2 c = 1
d = 1
h = 0.1
h1 = -0.1 h2 = 1
N = 31
M = 1
P = 1 e = 1
{Globals}
{End Globals}

SR003A equns.rtf is the following :
{Top model}

{Reservoirs} d/dt (R) = + dR INIT R = 0
d/dt (J) = + dJ INIT J = 0
d/dt (Z) = + dZ INIT Z = 0

{Flows}
dR = a * R * (1 - R) + b * J + e* Z + TT * (1 + k3 * R)
dJ = c * R + d * J + f* Z
dZ = R1+g * R

{Functions}
S = h * SQUAREPULSE (N, M) - h * SQUAREPULSE(N + P, M) + P * 0 a
= -1.8
b = -2 c = 1
d = 0.6 g = 0.8 e = 1
N = 31
M = 1 h = 2
P = 1
f = 0.1 k = 0 k1 = 0
k2 = 0.02
TT = k * RANDOM (k1,k2)

k3 = 1
V = 0
F1 = SIN (X1 * F2) X1 = time
F2 = 5
V1 = 1
R1 = V * S + V1 * S * F1
{Globals}
{End Globals}

M002 equations are the following:
{Top model}

{Reservoirs} d/dt (R) = + dR INIT R = 0
d/dt (J) = + dJ INIT J = 0
d/dt (Z) = + dZ INIT Z = 0

{Flows}
dR = a * R + b * J + e* Z dJ = c * R + d * J + f* Z dZ = S+g * R

{Functions}
S = h * SQUAREPULSE (N, M) - h * SQUAREPULSE(N + P, M)
a = -1.8 b = -2
c = 1
d = 0.6 g = 0.8 e = 1
N = 31
M = 1 h = 2
P = 1
f = 0.1
{Globals}
{End Globals}

A001btempnew
{Top model}

{Reservoirs} d/dt (R) = + dR INIT R = 0
d/dt (J) = + dJ INIT J = 0
d/dt (Z) = + dZ INIT Z = 0

{Flows}
dR = a * R * (1 - R) + b * J + TT * (1 + k3 * R)+ k4 * R^3 dJ = c * R + d * J +

```
f* Z
dZ = R1+g * R

{Functions}
S = h * SQUAREPULSE (N, M) {- h * SQUAREPULSE(N + P, M )} + P * 0
a = -1.8
b = -2 c = 1
d = 0.6 g = 0.8
N = 31
M = 1 h = 2
P = 1
f = 70 k = 0 k1 = 0
k2 = 0.02
TT = k * RANDOM (k1,k2)
k3 = 1
V = 0
F1 = SIN (X1 * F2) X1 = time
F2 = 5
V1 = 1
R1 = V * S + V1 * S * F1 k4 = 1
{Globals}
{End Globals}
```

#5. To do a fast Fourier transform of say R or J in we just press F on the graph of R or J versus time, altering the scaling if desired (in Graph/Axis Settings/Scales) so we can see the frequencies more clearly. Then if we want to we can, say, remove from or add to the frequencies displayed and then do an inverse FT if desired, bearing in mind of course, the comments in Harris (2007) or Prideaux (2000) or for a current overall assessment Pribam (2007). We can also read Bohm & Hiley (1993). The intention is not to assume a holographic interpretation of the brain but the FTs and inverse FTs condition us to think of possible holographic type interpretations of our model, though of course we are not obliged to do so in terms of what we have said so far.

Draaisma (2000) points out the apparent possible relevance of the homunculus paradox to any holographic model or assessment in several places in his book, in particular on pp 156-7, 178, 212-18, 226-8. First I will point out that this fine book contains much philosophical comment, which by its nature demands constructive consideration. Briefly - and of course there is much more to say - mentioning his comments on p218 it seems to me that since each part of a hologram includes a total (though perhaps unclear) image (Prideaux (2000),

Harris (2007)) so too if a truly complete image is said to actually include the homunculus, the homunculus will fade along with the image. So that an accurate representation which we give will include the homunculus.

For example if we take a somewhat Spinozan view of the universe, the total image of the moment of a person's consideration may well include the whole universe and this concept is possibly a good enough rough metaphysical expose - the metaphysics people often say things like this, it fits in well enough with the idea of the butterfly effect etc. Even though we aren't using it in the present study - and it is important to remember that we are not trying to write metaphysics and I feel that the homunculus concept tends to rather betray us into that field - it is probably not difficult to see how we could mathematically give a loose exposition through the work of Prideaux (2000) or even at a pinch Bohm (1993) - in fact a few weeks ago the idea was also independently brought to my mind during a popular lecture on Bohm at the Scientific and Medical Network, who often seem to deal with such rather outre ideas.

Given the result of the last paragraph, the model which we consider must inevitably be a rather faded and blurred hologram and we are downright lucky to have even that. In fact we are not likely to find the homunculus except as a blur somewhere in the diagram. So we can say "This is not like the story of the little Dutch girl holding a can with a picture on it of a little Dutch girl holding a can with a picture ..." to an undecipherable infinite regress - and you know even if it were, infinite regresses expressed as infinite series are not so bad in physics. But here there is, as far as I can see, no essential infinite regress anyway, and thus no homunculus problem implied or entailed. In physics it is almost a sad thing when a paradox is not there - we recall Zeno's paradox being said to be a precursor of the calculus and so on. But still, if a paradox is not there then it isn't, and that for the moment at least my own belief is that such an interpretation may quell homunculus paradox worries, even if we may be left with some even more inspired and oracular ideas in the painstakingly long run in considering, if we can and wish to, holograms.

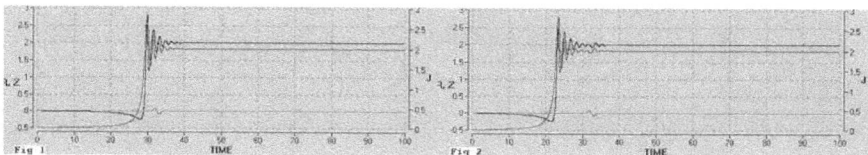

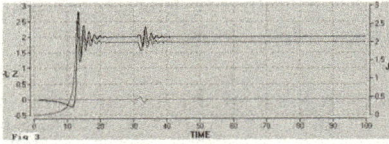
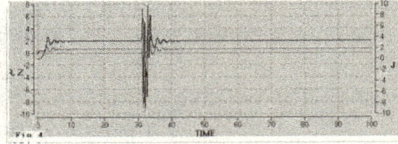
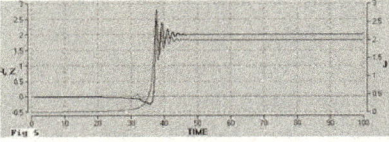
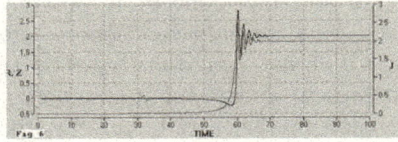
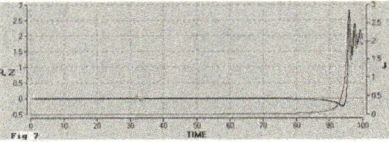
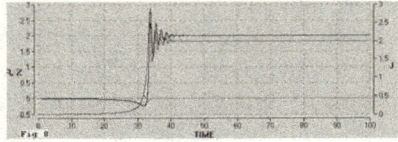
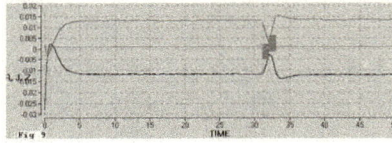

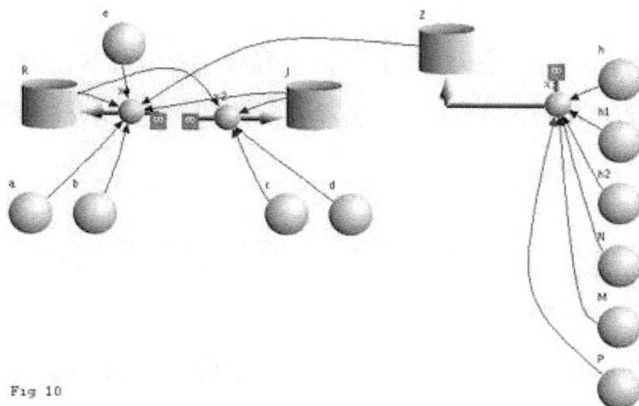

Fig 10

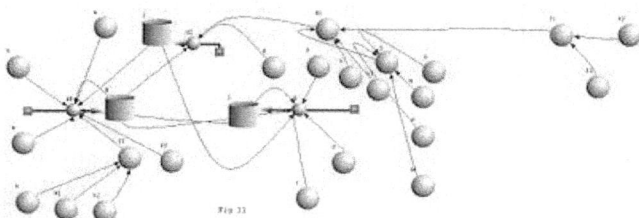

Fig 11

7.12 The Application of McTaggart's results to Consciousness Studies and Category Theory.

Abstract

It is explained how and why category theory may be used, together with reflexive monism, to explain the McTaggart paradox. This suggests the likelihood of even more applications in future.

Introduction to the Development of our Phenomenology

Ideas like "time" or "education" or "justice" or "mercy" are easy enough for most people to recognise, understand or use, but when it comes to the nub, quite hard to explicate or theoretically reconstruct. This tends to lead to nuances in interpretation which quite often are unnecessary or inappropriate in a particular context.

This fact comes out very clearly when relatively historically new concepts, like "electricity" are considered. Even today, it is sometimes said:

NA1: "In spite of everything we know about electricity, we do not know what electricity is." The idea being perpetrated is that whilst we know what charged particles are, and we know Maxwell's laws and other such properties of electricity, but nonetheless however much knowledge of this kind is gained we will never know what electricity is. But this is wrong because by knowing about charged particles, Maxwell's laws etc for the most part, we do know what electricity is. Of course we cannot sum up this extensive knowledge in a brief paragraph.

To clarify even further, in NA1 we could just as easily substitute the words "your nose" for "electricity" and make it NA1n. It is absolutely clear that we do know what "your nose" is and to say we don't is splitting hairs for most real purposes, philosophical details aside. Hard cases are likely to make bad law, here as elsewhere and we must rely on context and in the normal context, NA1n is absurd.

Referring now back to space and time, Buber (1959) pointed out 'A necessity I could not understand swept over me: I had to try again and again to imagine the edge of space, or its edgelessness, time with a beginning and an end or a time without beginning or end, and both were equally impossible, equally hopeless – yet there seemed to be only the choice between the one or the other absurdity'. The problem here is that when Buber tried to get down to philosophical details he just had not got the right stuff and relativity theory shows us that. There is really no certain reason, using relativity, why time or space would have a beginning or an end - philosophical problem solved.

Now we could say that Buber's confusion was caused by his acceptance of Newton's concept of space rather than Leibniz's. In Newton's world-view physical objects could exist by being in space, but space could exist even if devoid of any physical objects. In Leibniz's view, objects existed anyway and could touch one another, be separated by various distances and so on but space, per se, did not exist. This immediately resolved Buber's problem. One can solve such a problem by showing that it contains an untenable proposition. In this case the problem was not with space itself, but with Newton's conception of space. The answer was to accept Leibniz's more economical view, or simply to look for a consistent definition of space, which without relativity was hard to find.

McTaggart (1927) reasonably showed that in his context time showed a contradiction and he was right and logical to suggest that time did not exist, or is unreal. That was a sensible and economic view but slightly harder to develop than in Leibniz's case, where Leibniz had effectively inferred that space, per se, did not exist and was able to get quite a good theory for his era. But McTaggart's concern with time is in many ways very analogous with Buber's concern with space. Buber knew more or less what space was, but when he thought about it, it looked somehow spooky and unreal. Maybe we could say that that is "Angst". It is certainly a clear indicator that something needed to be done. Anyway, the same thing happened to McTaggart with time, and as we will pointed out here, just as Einstein resolved Buber's philosophical worry about space, so too category theory can up to a point resolve McTaggart's problem with time. But that of course does not give us the right to ignore McTaggart's problem just as relativity has shown we should not certainly not have ignored Buber's problem. Just as in a way we have all been erzatz Leibnizians, prior to Einstein, let us importantly try to avoid continuing the same line of error with McTaggart, whether or not a resolution of his problems is more of a serious mathematical and philosophical challenge than Einstein's resolution of Newton's problem was.

All the above should be roughly palatable to most people.

Time Travel in Theory and Practice

Quinton's work as interpreted by Norman Swartz and then by me.

Now Quinton (1962) describes a man living in England who when he falls asleep finds himself at a lakeside in a tropical setting. His experiences at the lakeside, unlike many dream-sequences, are as ordinary, as matter-of-fact, and as uneventful as are his everyday English experiences. Nothing dreamlike, fantastic, or wildly unlikely occurs in the tropical environment. The hero passes the day in the tropics and when he falls asleep there immediately finds himself in England. And these English and tropical experiences regularly alternate. Life in both locations appears equally real and Quinton has no reason to claim that objects in one reality have any spatial relations to those in the other, for no matter how hard he searches each reality he can never discover evidence of the other. That, at any rate, is Quinton's rather impractical concept-structure for the purpose of argument that two regions of space need not be connected and the structure has been considered again and again by philosophers since 1962. Swartz (2001) in his book, "Beyond Experience", reasonably claims that there is also a good case for taking it that all regions of time need not be connected either and we extend that notion here, as follows.

Quinton asks whether we can construct an analogous tale showing the possibility of there being two times. Surprisingly he argues that time, unlike space, is unitary, that there can be only one time. But the same dream argument can be applied to time as to space and, unlike Quinton claims, but as philosophers subsequent to Quinton sometimes correctly claim (Swartz, 2001), there is no reason why the two regions of time need be connected either! This is because when Quinton's hero, according to Quinton, keeps a diary 'At the beginning of day 1 in England I write down in order all the lakeside events I can remember. On day 2 in England I cannot remember whether the events of day 1 follow or preceded the lakeside events in the list. But the list will be there to settle the matter and I can, of course, remember when I compiled it'. How does Quinton's hero know that he is writing down memories (of past events) and not precognitions (of future events)? I'll specifically quote Swartz here ' No mental phenomena carry with them an identifying mark of memoryhood. Some of the things we think we recall never happened at all, yet introspectively these thoughts (images, beliefs) are indistinguishable from genuine memories. Merely being memorable or, more exactly, having the felt quality of a memory is no guarantee of the truth of that which has that quality. Our mental faculties can be faulty or deceived. We can believe that we remember events and it turns out that these events did not occur, and similarly we can fail to remember events that did occur. That a thought or mental image is really of a past occurrence and not perhaps a precognition of a future one is guaranteed by nothing in the thought or the image itself. The manner in which memories and precognitions present themselves to consciousness seems to be all of a piece. To learn that a memory-like thought is really a memory and not a precognition we must depend ultimately on objective criteria and more exactly on physical criteria.' Now that is, very briefly, Swartz's approximate argument and whilst there may be philosophical problems as to what we can reasonably consider as a physical criterion and what as a mental criterion, that is Swartz's argument as it stands and it could be amended further if we were desirous to do so. As Paul Churchland (1985) points out, for instance, precognition is occasionally occasionally cited in favor of dualism. Be that as it may, the argument does show that we cannot therefore logically infer that there is one unitary time but not one unitary space and Quinton's argument as amended shows that neither one unitary time nor one unitary space is required. We could say that modern string theory would presumably not object to that idea anyway but it would be naively accepting to use string theory simply on that basis, particularly after the work of Woit (2006), Smolin (2006) and others and we most certainly cannot accept all the wild and woolly dictates of string theory without useful qualifications. Indeed we may not well wish to support string theory at all.

How to relate these views to Cognitive Science

At the moment, perhaps the most popular philosophical explanations for cognitive science and mental phenomenology are dualist and of a physicalist or functionalist nature,. However the currently most plausible explanation may be Velmans's reflexive monism (2000 et al).

Jeffrey Gray (2004) pointed out that functionalism will not work, as conscious states cannot be identical to functional states (Note 1). He cited synaesthesia as synaesthetes can apparently instantiate the same conscious state type without instantiating the same functional state type. As he pointed out to me during a meeting in December 2003, he only needed to find one such type of case, or indeed technically only one case, to disprove functionalism per se. Functionalists can claim, of course, that synaesthetes do not experience the same qualia in the two functional states or they could adopt a supervenient functionalism or something else. There is an element of question begging to such contrary arguments and whilst there are several such possibilities, then reasonably assuming Gray is correct (Note 1), the fabric of functionalism is wearing thin to the extent that simply because of Gray's work it is becoming more practical to consider, for example, Velmans's reflexive monism instead.

Velmans (2000 et al) has given many cogent reasons in favour of reflexive monism anyway and reflexive monism is not thrown over by synaesthesia considerations.

Kihlstrom (2002), (who by and large supports Velmans's views), makes a number of points about the placebo effect. Kihlstrom mentions, with examples, the fact that placebo results must be taken very seriously. Indeed it has been said recently (Hislop, 2006) that the placebo effect helps the UK National Health Service to keep alive, partly due to the ability of the NHS to refer its clients to 'alternative' (and often absurd) practitioners. Kihlstrom himself suggests that the placebo effect may work by allowing conscious beliefs to alter bodily states and implies that work like Velmans's may provide part of the clues to this. My own experience in altering my own pulse rates and so on by using biofeedback suggests that whilst such factors undoubtedly exist and may indeed be highly beneficial (Note 2), that there is much more hard science required if we want to gain a thorough understanding of such effects though Velmans's work may lead to steps in the right direction. So there are results that scientific experience and not just probable illness remission rates but immediate experiments with a pulse rate monitor make clear. Clearly physicalism sounds as if it ought to be a worthwhile and sensible explanation of mental processes but reflexive monism seems to leave more opportunities for investigative progress.

How to relieve an Angst-ridden world of the problems of McTaggart

Following the theme of the introductory section, we do want to relieve everyone of the Angst that McTaggart's ideas seem ineluctably to lead to, just as Einstein relieved the world of Buber's Angst, and as we mentioned in the section on Quinton's work, a sort of 'precognition' might be definable in terms of our new and less Angst-ridden description of physics. Now 'precognition' or if we want to call it that 'presentiment' is a very popular (or populist) notion but one that is, for very good reasons a notion which is almost anathema to current highly respectable physics literature (Note 3), though not to the philosophy literature to the same extent. Precognition served as a useful tool for Swartz in his comments on Kant. So does precognition fit neatly into our program, or perhaps more to the point, can the concept of precognition enlighten us as to the possible direction of any programme? Can precognition at least serve as a torchbearer during a relay on the long road to enlightenment?

Time Travel in Theory and Practice

As I mentioned in the last section, Paul Churchland (1985) pointed out that, for instance, precognition is occasionally cited in favour of dualism. Now today when we think of dualism we may want to cast the net somewhat wider than that if we can, and indeed the obvious approach to consider may well be the reflexive model of Velmans (2000, 2006), and that or something like that seems worth considering. In the sections below we attempt to follows Velmans's approach.

Much of the comment in this section is mine, and - at Velmans's request - should not be attributed specifically to Velmans except for comments which actually occur specifically in his own work as it is important that he should spend his time answering objections to what he has actually said himself wherever possible.

As Velmans (2006) points out cogently, ideas like his about the spatially extended nature of the experienced phenomenal world 'fit in with common sense and common experience and they will come as no surprise to those versed in European phenomenology have many theoretical antecedents ... for example in the work of Berkeley, Kant, and Whitehead, the neutral monism of James, Mach, and Russell, and the scientific writings of Köhler and Pribram'.

Now to test this to see how it works out: I won't repeat the exposition of Velmans (2006), (which is also referred to in Velmans (2000)), but Figure 1 of Velmans (2006) describes dualism, Figure 2 describes reductionism and Figure 3 describes the reflexive model. In the reflexive model, and there is a dynamic interaction between the observer and the observed.

In dualism (Figure 1) a percept simply occurs in the mind. Precognition can be incorporated by leaving it open as to what goes on in the mind of the individual, for example they could be mentally incompetent during an imagined or real precognition experience.

In reductionism (Figure 2) it is (normally) argued that the percept in the brain is simply a function of the brain and whilst it is separate from the brain, it can still be described in a way (straightforward in principle scientifically) in terms of what goes on in the brain. So the same position arises for precognition. Neuroscience can show that it may occur in the brain, for example due to mental incompetence or possibly simple lack of understanding or naive behaviour.

Velmans justifies his reflexive monism (Figure 3) and points out that it has falsifiability as well as reasonably claiming that physicalism lacks it. Also. he takes pains to point out that he is not an epiphenomalist. These matters are covered in some detail in Velmans (2000, 2002,2002a).

Before we go any further I want to make two points.

1. Velmans M., (2001a) states 'Definitions need not be final for research to get under way. It is enough that, for given investigative purposes, definitions are sufficiently similar for different investigators to be able to agree that they are investigating the same thing. As science begins to unravel the causes of consciousness ... and so on'.
2. Velmans (2001) says, whilst describing a discussion with Dennett 'it does not follow that conscious qualia have no useful place in "first-person" accounts, nor that they do not exist.' This would seem to be a position retained to the present time in using the reflexive model and so we may still use qualia to consider such reflexive models at least for explanatory purposes.

On page 240 Velmans (2000) suggests that attentional processing is closely related to consciousness so that the contents of consciousness become a kind of "psychological present" which is immediately accessible for processing, including the relevant sections for current action, suggesting that input analysis becomes conscious when its products are being disseminated. In my view, consciousness of the past may be through some kind of long term autobiographical memory and consciousness of the future by some form of speculated brain-computer interaction or through a brain mechanism related to planning and evolving which could contain a series of explicit or implicit rules and interactions, such as the saw "red sky in the morning, shepherd's warning, red sky at night, shepherd's delight". I am currently exploring and trying to more clearly delineate other possibilities of 'consciousness of the future' in my blog at http://ttjohn.blogspot.com/ and of course the largely populist idea of 'precognition', which has some theoretical philosophical force (e.g. Churchland, 1985) still occurs. Much of what is perceived by the brain is known not to be overtly conscious. In some cases attentional processes are partly dissociated from consciousness, and in such cases perception and how it could be occurring can often be measured or noted, as in blindsight and dreams. So we may be already getting somewhere near where we may be able to consider empirical problems and not conceptual problems, indeed 'easy' problems rather than 'hard' problems. And indeed Velmans says essentially that on p241.

If we wished to describe the consciousness of one individual at one time, for example, we could have a narrow attentional section at about that time, an at least partly autobiographical past section and a 'possibly speculative' future section, not just a single dot. The unconscious section would be far more complex and certainly we do not know it all at this time. We are not talking here about the physical state correlates.

By the time he gets to p243 Velmans in fact points out (indeed re-iterates as he first said it in Chapter 1) that each individual has his own consciousness. Thus each individual has his own past- present-future and to conflate each and every individual's past-present-future is not a recipe for accurate physics! In short, the space time cones which relativity uses are actually not good enough to use for a number of people interacting or just existing. Certainly we can treat the people as if they are inanimate objects (who can for instance signal mechanically one to the other) but we must not confuse this with an accurate physical (or NCC) picture involving their own thinking. In a subsequent section I will try to indicate how this may be done, however.

It certainly is possible to lay down the mathematics but it does probably require each individual to have at least their own past-present-future (PPF). Clearly there could be similar looking regions in various people's PPF and these could be possibly approximated to carefully using category theory. An analogy might be the common symmetry structures of models of benzene with and/or without sidechains, for the purpose of bulk determinations like UV spectroscopy. But Velmans carefully distinguishes the cat (of his example) as seen in the brain (which for the subject is the 'real' cat) from the NCC which occurs in the brain. It seems to me that the 'real' cat contains the qualia whereas the NCC are an objective map. We can deal with either, and the NCC will, we presume, be relatively observer-independent but the 'real' cat as observed will vary a lot from subject to subject.

So we now have the added complication of TWO plans or maps.- the very likely relatively invariant NCC map and the subject-'real' map which will alter from subject to subject., This subject-'real' map could possibly be approximated to however, in the way just described and then may seem more like a NCC map.

I am not sure that Velmans's wave/particle duality comparison (on p250) can be pursued as other than an analogy and I think the whole reflective monism description is best described in category theory terms rather than initially as quantum mechanical terms, as the effect may well not even need a quantum mechanical explanation. Effects such as the Kanisza triangle and other optical effects such as the well known line drawings of cubes with a corner which seem to stand in wards or outwards from the page almost at random, may be helpful in describing what is happening. These effects are so far more commonly personally observed rather than automatically through NCCs (though the effects could be simulated by a computer, they would generally have to be very contrived for simulation). It seems to me, also, that the 'unconscious' effects (page 253) which Velmans considers are allied to the conscious subject-'real' world may have to be allied to the subject-'real' map though would clearly not appear in the subject consciousness, as they are unconscious. So now we have at least the following items to consider.

1. The cat as consciously perceived by S (in Velmans (2000), fig 6.3 or Velmans (2006), fig 3)
2. The NCC of S (in Velmans fig 6.3)
3. A mapping of the cat in #1 (One such mapping would literally be the picture of the cat in figure 6.3)
4. A mapping of the NCC in #2
5 A mapping of the conscious and unconscious perceptions of the subject S whilst the subject observes the cat.
It all sounds worse than the Danzig corridor and the 'corridor within a corridor' but it probably cannot be helped and the functors need to be considered in detail also.
We also have
6. The cat as it is generally understood objectively
7. A mapping of the cat in #6.

and mappings being what they are, there are probably others as well. We may wish to assume that
#4 does not need a further mapping of "the NCC as objectively understood" and so on.

On page 253 Velmans gives a plausible enough 3 step solution of his own 'causal paradox'. Importantly he points out (page 277) that in his scheme 'consciousness is the creator of subjective entities' . We are quite happy that Velmans insists on the inclusion of a large number of subjects as individual parts of his overall pattern though he himself feels that some philosophers may find difficulties with it. Mathematically it looks no worse in principle than accounting for individual particles of differing structure in a Boltzmann gas. Though we may for example have to have two categories (say, MacA and MacB) to account for and this fact in itself concretises the system as a possibly conscious one, as it can become conscious once it transcends punctal time !

Time Travel in Theory and Practice

Without such transcendence, #1 and #3 above will not contain adequate factors to describe human consciousness. We know that humans certainly do not observe time as punctal and any physical representation of consciousness, such as the spacetime continuum of Einstein, is thus suspect in terms of 'brute fact' or simply carried out observations! Psychologists today seem to want to exhibit a form of scotomisation and refer to 'human time' as 'specious time' but it is still a clear attribute of consciousness. It may be that other factors may later have to be added, but these have not yet apparent and the scheme can be altered to fit them. We really want to keep it as simple as we can.

As Velmans points out on page 278, 'if all humans were removed from the earth, only a mechanical earth without consciousness would remain'. It is essential that an earth with humans should have a different description to an earth with humans, or the physics would be wrong. We could of course generalise the above statement, if we needed to refer to 'an earth without life in any form'.

So we certainly include factors here which should make the physics better than simple Einsteinian physics, include human beings and sort out McTaggart's paradox. That is not to say that we have gone far enough yet but it looks like a fairer starting model than we presently have with the efforts of Einstein and more contemporary thinkers. Velmans ends his book by fairly quoting an apothegm of Jung 'man is indispensable for the completion of creation - in fact he is the second creator of the world, who has given the world its objective existence' and we know that Velmans's basic ideas tend to agree with those of Sri Aurobindo (Velmans (2000), p167).

Crucially, unlike the variants of physicalism and functionalism defended by Torrance, Van Gulick, and Chrisley & Sloman, the reflexive monism of Velmans also conforms closely to the evidence of first-person experience and it thus suits the present category-theoretic formalism.

We want to stick, where possible, to the conventions of Lawvere (2005) and in due course may be making use of the computer program (Graves, 2002) for some of the mathematics. This of course does not foreshadow any extreme philosophical representations, such as the 20th Century notion that the human brain can be (or can't be) replicated in a computer.

Some ongoing experiments and further work are given in my blog, http://ttjohn.blogspot.com/ , which also gives details of my website, email address and egroup.

I wish to thank many people but in particular Max Velmans, Basil Hiley, Christopher Isham, Allan Hobson and the late Jeffrey Gray for helpful discussions concerning this manuscript.

References

Brown R., (2006). For example Brown R., Paton R., Porter T., "Categorical language and hierarchical models for cell systems", to appear in Computing in Cells and Tissues - Perspectives and Tools of Thought', Springer Series on Natural Computing. More preprints of material like (Brown, 2006) such as 05.13 are on Brown's home page at http://www.bangor.ac.uk/~mas010/ or at http://www.informatics.bangor.ac.uk/public/mathematics/research/preprints/

Buber, M., (1959) Between Man and Man [1947], trans. Ronald Gregor Smith, Beacon Press, Boston.

Churchland P., (1985)Matter and Consciousness: A Contemporary Introduction to the Philosophy of Mind, The MIT Press, Cambridge, Massachusetts, 1985; http://www.msu.edu/~marianaj/church.htm

Ehresmann A., and Vanbremeersch, j-p. (1999) "Memory Evolutive Systems", http://cogprints.org/921/ ; also see C., http://perso.wanadoo.fr/vbm-ehr/Ang/W24A5.htm , http://perso.wanadoo.fr/vbm-ehr/Ang/W208.htm

Graves M., Rosebrugh R., (2002) "Graphical Database for Category Theory", Version 2.0, Jeremy Bradbury, Ian Rutherford, July 1, 2002

Gray, J. (2004), "Consciousness: Creeping up on the Hard Problem", OUP

Hislop I., ed (2006) "Private Eye", Vol. 1160, page 11

Kihlstrom J., (2002), Journal of Consciousness Studies, 9(11), 30-34.

Lawvere C.W., Schanuel S.H., (2005) "Conceptual Mathematics", Cambridge. and see Note 4.

McTaggart J M E, (1927), "The Nature of Existence" , Vol. II. Cambridge: Cambridge University Press.

Quinton, Anthony, "Spaces and Times", in Philosophy 37, no. 140 (Apr. 1962),

130-47.

Savitt S., (2002) "Being and Becoming in Modern Physics", The Stanford Encyclopedia of Philosophy (Spring 2002 Edition), Edward N. Zalta (ed.),

Smolin L., (2006) "The Trouble With Physics : The Rise of String Theory, The Fall of a Science, and What Comes Next", Houghton Mifflin, ISBN: 0618551050

Sudsuang R. , Chentanez V., Veluvan K., (1991) "Effect of Buddhist Meditation on Serum Cortisol and Total Protein Levels, Blood Pressure, Pulse Rate, Lung Volume and Reaction Time" Physiology & Behavior, Vol. 50, pp. 543-548

Swartz N., 2001. "Beyond Experience: Metaphysical Theories and Philosophical Constraints", Chapter 8, http://www.sfu.ca/philosophy/beyond_experience

Tovey D., ed. (2006), "Clinical Evidence", www.clinicalevidence.com, The BMJ Publishing Group, BMA House, Tavistock Square, London WC1H 9JP

Uchii S., (2003) http://www.bun.kyoto-u.ac.jp/~suchii/mctaggart.html

Velmans M., (2000) Understanding Consciousness, London: Routledge/ Psychology Press.; and Velmans M., (2002), "How could conscious Experience affect Brains ?", Imprint-Academic, UK. which includes criticisms and appraisals of Velmans's work and his answers to them.

Velmans M., (2001) "Heterophenomenogy versus critical phenomenology: a dialogue with Dan Dennett", http://cogprints.org/1795/index.html

Velmans M., (2001a) "A natural account of phenomenal consciousness", Communication and Cognition

Velmans M., (2002) "Making sense of causal interactions between consciousness and brain". Journal of Consciousness Studies, 9(11), pp. 69-95.

Velmans M., (2002a) "How could conscious experiences affect brains?" Journal of Consciousness Studies, 9(11), pp.3-29

Velmans M., (2006) "Where experiences are: dualist, physicalist, enactive and reflexive accounts of phenomenal consciousness", cogprints.org/4742/

Varela F.J. , (2000) "Naturalizing Phenomenology: Issues in Contemporary Phenomenology and Cognitive Science" Edited by J. Petitot, F. J. Varela, B. Pachoud and J-M. Roy, Stanford University Press, Stanford Chapter 9, pp.266-329

Varela F.J. , (2000a) "The Specious Present : A Neurophenomenology of Time Consciousness Francisco" in: J.Petitot, F.J.Varela, J.-M. Roy, B.Pachoud "Naturalizing Phenomenology: Issues in Contemporary Phenomenology and Cognitive Science", Stanford University Press, Stanford

Woit P., (2006),"Not Even Wrong: The Failure of String Theory and the Continuing Challenge to Unify the Laws of Physics", Jonathan Cape, ISBN: 0224076051

Notes

1. My results and/or comments on synaesthesia suggest, on the basis of a considerable number of interviews with, readings of records of opinions of and conversations with, and actual tests carried out by me on some of the synaesthetes considered by Gray, and others, that even after MRI investigations, the precise scientific standing of the phenomenon is still unclear. Only two reasonably well authenticated cases of 'pop out' for normal subjects are known, for example. Generally speaking, most people who have studied the phenomenon agree that there is some sort of real effect but the exact nature of the effect needs more work.

2. There are references as to the positive effect of the mind on bodily effects in Kihlstrom's paper and indeed throughout the literature, for example Sudsuuang (1991) and many other papers. But nota bene that at the time of writing this paper, no details in www.clinicalevidence.com (Tovey, 2006) seemed to show that biofeedback actually cured illness or ameliorated pain on a relatively certain basis, by statistically adequate trials. But the physical effect mentioned in the main text here certainly exists and can be measured.

3. This short paper will not intend to outline the reasons for this - as an analogy, think of previous "discoveries" of "low temperature nuclear fusion" to obtain the kind of problems which could be needlessly caused to the establishment.

4. At times the conventions of "Basic Category Theory for Computer Scientists", Pierce B.C., (1991) may also be needed. But I am trying to keep it all as simple as I can and obviously other textbooks by authors such as Barr and Wells, Awodey, McLarty and MacLane may be needed at times.

Time Travel in Theory and Practice

Epilogue

This book does not contain emphasis on possible temporal paradoxes nor upon the possibility that the broad discovery of any new and strange facts about time may give to the world important societal and similar changes. So many people have already discussed temporal paradoxes and the purpose of this book is to advance the subject, not merely to discuss it. But certainly we continually refer to these matters as and when we need to, and they are continually a subject of thought.

An obvious query is: In what way could further developments of these methods, apparatuses and theses we have described here alter the world if they are developed. We can ask this hypothetically, or as a preliminary to their factual adoption.

There are many answers. First consider the development to date of space technology, to give us an inkling of any potential future to these ideas. For example the meaning of the presence of humans in the universe and possibly in any as yet undiscovered addtional areas has widened considerably, since the days of Copernicus through Jules Verne and now to the idea of enormous new vistas of discovery.

Like Superman or the man on the flying trapeze, we now fly through the air with the greatest of ease, not just into the air but through the entire universe and possibly through even more universes yet to be confirmed. But this is not easy for individuals. We are not all behaving like comic characters such as superman, in fact we normally can travel through the air in large expensive metal and plastic boxes, and may well worry whether we will ruin the earth by doing this so much. And as for getting to the moon and the stars, these are still too expensive even for most countries and we so far seemed to have gained most perhaps by non physical new information and by improved communications. Most people cannot even afford their own private flying car even after 100 years of ground transportation. And such practical possibilities as mining on the moon may only be realised in circumstances of dire need. It's all really not like most people would have expected, even 100 years ago.

Time Travel in Theory and Practice

At the moment here we are merely humbly saying that the mind is not yet fully understood. Thus any proposed future scenario should probably be very different to the traditional science fiction - and, it must be admitted - common layman's viewpoint. And that view so often boils down to presupposing - if there is to be any great change - something like a simplified Tralfamadorian perspective as given in Vonnegut's "Slaughterhouse 5" as explained by Nick Huggett in a learned reference given in the Introduction. That simply involves considering our entire lifespan as like a long worm and other homely manifestations like that. That is probably all wrong, as a simple overview. Now we have enlisted the 'second blind man of Indostan' referred to in Chapter 1, whereas earlier work mainly centred around one of these gentlemen. We need at least to consider two blind men of Indostan.

We also need to get away from the very basic 'block time' idea which when you think of it, is totally alien to our own mental thoughts or feelings. The mystics, the Buddhists the dreamers and so on, may only form a tiny part of the whole world but it is all real, and so cannot simply be ignored as 'folk beliefs'. And the 'specious present' is real, because to us it really is there.

We think of the simple humble old idea of the earth being mostly all there is to life, with the heavens and the rest being more the province of a perhaps unreal 'god' or 'gods' and compare it with the enormous amount of supposed information which modern science gives us about a large universe, which scientists would have us believe is real. Yet we still, basically, accept this very humble 'earth' as being our real environment, much more real if you like than pictures of the 'moon' and of 'mars'. Of course if for some reason we actually go on a space ship, that will merely become almost automatically simply an extended version of our reality on earth. And even to push the matter further, if we were to end up living on Mars, the TV program 'Life on Mars' would seem to me to be a roughly adequate example of what we might expect. Certainly the same program seemed to give a roughly adequate idea of the Salford, Lancashire police in the mid 20th century. I spent too much time in Salford, Lancashire to doubt that. So we have not left behind 'folk beliefs', they are still part of our mental package today.

But given the above, so too are the 'folk beliefs' of mystics, Buddhists, dreamers and all the rest. We take it that California and Timbuctoo both exist in some sense, even if we have not been there, and similarly the 'folk beliefs' of mystics, Buddhists, dreamers cannot be viewed sensibly with complete scepticism. Potential examples are implied in Chapter 7.01 for example, as these beliefs are at least as real as an atom of copper in the nose of a statue.

Time Travel in Theory and Practice

They are certainly there whether we conceive them as 'real' qua 'scientific fact' or not. And we could bog ourselves down by taking a contrarian standpoint and asking if science is really 'there' in the same sense. But rather we must remember we are aiming to project likely future achievements, just as Einstein tried to use simple visual observations on railway trains to deduce special relativity. And we must be careful not to conflate our lucubrations with 20th century physics, just as Einstein was careful not to conflate special relativity with Newtonian physics.

All this is not easy and it has to be remembered, for example, that without Noether's theorem we possibly would not have general relativity or modern particle physics. Of course Noether's theorem may have been discovered at some time though it might well have been simply ignored or forgotten. We must remember that Einstein only achieved a plausible general relativity theory by having the acquaintance of Emmy Noether. So there is still a lot to learn, and much of it is unlikely to occur as redevised 19th century mathematics, or indeed as mathematics at all.

Alphabetical Index

The alphabetical index above refers to areas of interest regarding the indexed word, rather than to specific pages of indexed word use.

www.ingramcontent.com/pod-product-compliance
Lightning Source LLC
Chambersburg PA
CBHW022052210326
41519CB00054B/316